PISTON

Made from iron of special formula for pistons and given the same heat treatment as our cylinders and ground to perfect round on the grinder.

WATER JACKET

GRAY Motors have liberal and ample water spaces.

COMBINATION RELIEF COCK AND PRIMING CUP

PISTON RINGS

Made from special piston ring metal with both edges and face ground and pinned to prevent turning. In spite of the fact that we grind all pistons, rings and cylinders, we also lap these pistons into the cylinders as well, making the most perfect fitting piston that mechanical skill can produce.

PISTON PIN

Hollow high carbon steel, case hardened and ground.

PISTON PIN BUSHINGS

All pistons on GRAY motors have bronze piston pin bushings.

CONNECTING ROD

Drop forging, high carbon steel.

COUNTERWEIGHTS

Made of malleable iron fitted on crank shaft with machine fit, dowelled and riveted.
The fitting of the counter weight on the crank shaft is a very important thing and it cost the GRAY Motor Company many thousands of dollars to find out how to do it right. The centrifugal action here is very difficult to overcome.

COUPLINGS

Made on automatic machine absolutely true and perfect. Not a cheap sleeve coupling, but the best and most expensive type of coupling it is possible to build, keyed to the shaft with perfect fitting keys.

CRANK SHAFT

Made of high carbon steel, over 120,000 pounds tensile strength from a private formula of our own for heat treatment, which gives 40 per cent stronger crank shaft than the same steel without the treatment.
Strong, accurately ground, perfectly fitted crank shafts are features of GRAY motors.

THRUST BEARINGS

We use the highest type of thrust bearings, not the ordinary cheap bearings frequently used in marine motors.

MAIN BEARINGS

Note carefully the extreme length of the main bearings on GRAY motors.
These bearings are not only very long, but a perfect fit of the crank shafts in them, making the life of the GRAY motor depend almost entirely on the attention you give to your lubrication.
They last practically forever with proper lubrication.

DRAIN COCK FOR BASE

Engines.

Ex Libris

MANCHESTER YACHT CLUB

Given by

_____

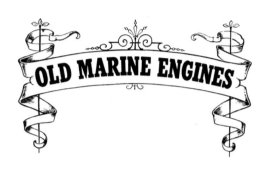

OLD MARINE ENGINES

The World of the
ONE-LUNGER

STAN GRAYSON

INTERNATIONAL MARINE PUBLISHING COMPANY
Camden, Maine

All photographs were taken by the author, except where indicated otherwise.

© 1982 by International Marine Publishing Company

Typeset by Journal Publications, Camden, Maine
Printed and bound by The Alpine Press, Stoughton, Massachusetts

Published by International Marine Publishing Company
21 Elm Street, Camden, Maine 04843
(207) 236-4342

Library of Congress Cataloging in Publication Data
Grayson, Stan, 1945-
  Old marine engines.
  Includes index.
  1. Marine engines.   I. Title. II. Title: One-lunger.
VM731.G72 1982   623.8'723   82-80402
ISBN 0-87742-155-2

For Grandma Jennie

Contents

Preface

One clear spring day in Bridgewater, Nova Scotia, when the sky was bright blue and the wind just ruffled the green surface of the La Have River, I stepped across a threshold. With me was engineer Stan Forbes, and, together, we were confronted by a big yellow diesel engine. A methodical workman was hooking up fuel and exhaust lines. "Look oot, boys," he chortled, "soon I'm going to make a big noise in here."

We were standing in the testing house of Acadia Gas Engines. Although it was destined to be closed within a few months, the testing house was, at the time of my visit, still being used as it had been for more than a half-century. Even so, it had already been more than a decade since Acadia regularly tried out its famous two-cycle engines here. Once, when the company was actively manufacturing such machines, four at a time might have bellowed away in the testing house, under the careful eye of a mechanic. Now that is all over.

"See here," said Forbes, pointing out a dust-covered chart upon the wall. "Here's what we used to refer to so as to get the right load on each size engine." The lettering on the chart was faded brown and indistinct. The glass covering the little document was coated with grime.

"It still seems a little strange," said Stan Forbes, who had been for many years an engineer at Acadia but who is now retired, "to see a diesel engine in here." A few minutes later, as we walked along the river where schooners once had docked, we heard the diesel come to life with a knocking clatter.

The two-cycle marine engine, the classic one-lunger, belonged to a world that has ceased to exist. It was a product of the great move toward mechanization and away from a centuries-old reliance on the craft of sail. Gas engines spawned a new craft. For a quarter of a century and more, these engines represented a way of life for their owners. They were part of a working world in which, to survive, a fisherman had to know how to fix things by himself. He had to know how things worked. Mostly, these fishermen lived in scattered harbors and outports all along the coasts. They were able men, and they got by. Their two-cycle engines were as much a part of daily life as horses or boats.

Today, there are not many people left who actually lived that life, or who recall in much detail those who did. Most of those involved with old gas engines today are dedicated amateur historians or collectors who value the old machines as fascinating antiques. One cloudy day in 1974, I first became aware that some people liked old engines enough to save them. I had driven to Connecticut to photograph a 1914 Trumbull cyclecar. In the barn where he kept the car, the owner also had a number of large stationary gas engines, each one restored, painted, pinstriped, and so mounted that it could be trailered to the meets where enthusiasts gather to marvel at each others' machinery.

The man who owned the Trumbull and the engines was named George Clark, and we became friends. Years later, George sold off the Trumbull, but he kept his old gas engines. His fascination with them has, if anything, grown. The last time I saw him, he had just acquired a valuable and impressive collection of *The Scientific American,* pre-1900. Each night before going to sleep, he told me, he scours their yellowed pages for articles about gas engine development. Occasionally, he discovers a device so fascinating that he feels impelled to re-create it. Because he is a machinist of supreme skill, he usually succeeds.

There are engine collectors throughout the world who share George's enthusiasm. In their workshops, basements, and barns are kept alive the names of the old engine companies and visions of a time when the world was a less cluttered and, perhaps, less complex place. Because many of these enthusiasts wanted to know more about early marine engines, they wrote to International Marine Publishing Company, asking for information or advice. That is partly how this book happened. Roger Taylor, the company's president, decided to move ahead with such a project, and he asked me if I would like to be the author.

In the autumn of 1980, I went out looking for those who had actually lived the story of the early marine engine. Most of them, I knew, had already, as naval architect William Atkin once wrote, passed "down a sun-lighted channel to the sea of no return." I did not know if I could actually find any who survived, but I wanted very much to do so. I wanted to hear their stories before it was too late and see the world as they had seen it. I found them in a scattering of small towns from Maryland to Maine, from New Jersey to Nova Scotia, and they shared with me their recollections of boats and men and motors they once had known.

They spoke of springtime days in Newfoundland and the onset of scalloping time on Cape Cod. They spoke of dawns when they awoke to the lively bark of unmuffled two-cycle marine engines echoing off harborside buildings and throughout a waking village. The memory of that sound, of that steady pock-pock-pock-pock, remains strong in them even now. It was as if they could hear it still, as if all the intervening years were only yesterday.

Stan Grayson
Marblehead, Massachusetts

Acknowledgments

Before it was determined that this book could, in fact, be written, several people were of particular help. One was Bill Smith, of New Jersey, as careful a rebuilder of old marine engines as he is of Rolls-Royce motorcars. Dick Day and his wife, Barbara, shared their enthusiasm for old engines with me and read a portion of the manuscript. Naval architect Fenwick Williams dug into his files and memories and came up with both information and encouragement. In Canada, Lloyd Conrad at The Lunenburg Foundry and Don Boutilier at ABCO-Acadia took time from busy schedules to assist me in every possible way. A special thanks to three men in Maine: Bob Rice, Steve Lang, and Joe Torrey; to Peter Vermilya at Mystic Seaport and Gerald Lestz, editor of *The Gas Engine Magazine*. Mrs. Eleanore Clise of the Geneva (New York) Historical Society, Mrs. Alice Austin, Mrs. Hilda Weston, Morris F. Glenn, Owen Barcomb, and Ben Brewster all helped at various stages. So did the late Jim Bradley of the Detroit Public Library, his assistant, Margaret Butzu, Sohei Hohri of the New York Yacht Club Library, and Joe Gribbins. Thanks, too, to Bob Merriam, Gerald Smith, Edward Welles, Sr., and Dudley Davidson, enthusiasts all, and to my old friend, Karl Aronson, as fine a natural mechanic as I've ever known, who taught me some basic secrets about two-cycle engines, many years ago.

OLD MARINE ENGINES

"Now, then," he said, "you prime her, and I will start her. Huw, my little one, you are helping to make history. Hold on now . . . ."

And now it fired, and fired again, and Owen turned no longer but pulled the handle clear, and looked as though to make it run by his will. Quick as quick the firing came until it was in a storm of firing, shaking the place under my feet, making me clench my jaws.

The engine was going. After years, it was going.

Richard Llewellyn, *How Green Was My Valley*

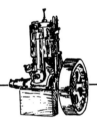

1

The Coming of Power

IN THE LAST DECADES of the 20th century, one must prod oneself to realize that, at the century's beginning, people could be as thrilled by their first sight of a motorboat as a later generation would be by the footprint of a man on the moon. In 1902, after he had successfully powered his way around the tree-lined harbor in Chester, Nova Scotia, inventor Frank Hawboldt was carried triumphantly away from the wharf on the shoulders of townsfolk whose only experience of small craft had involved sails and oars. Hawboldt had built a two-cycle engine in his barn and, overnight, had changed a whole town's outlook on life. Naval architect Weston Farmer saw his first engine when he was four years old in 1907. "The novelty of seeing a boat 'rowed' by an engine will stay with my wonderments until I die," he wrote.

Few devices have so altered the way we live, in so short a time, as the internal combustion engine. That it had the potential to do so was recognized within a relatively few years of the Civil War. When he successfully demonstrated an engine of his own design in 1872, George Bailey Brayton prompted so much interest that special congressional hearings were held about what the gas engine might portend. "It may someday prove more revolu-

The first Hawboldt, a make-and-break two-cycle marine engine. Hose goes from water pump to water jacket. (Courtesy Chris Hawes, Heritage Foundation)

tionary in the development of human society," said the congressmen, "than the invention of the wheel, the use of metals, or the steam engine." Brayton's engine was seized upon by inventor John Holland, who saw it as a way to build a practical submarine. The engine was improved by another inventor named George Selden. In 1877, he installed the machine in a horseless carriage and for a number of years owned a patent on the automobile.

Gas engines took North America by storm. When he published his 1910 edition of *Gas, Gasolene and Oil Engines*, Gardner Hiscox estimated there were then "about 10,000 manufacturers building gas, gasolene and oil engines in the United States." They sprang into existence as part of the profound revolution in technology that Congress foresaw in 1872. Internal combustion engines found their way into most aspects of life: industry, agriculture, and transportation on land, sea, and air. Few of the many manufacturers in those early, optimistic days guessed that the market could not possibly sustain them all for long. Instead, there was a craze to produce, and engine builders sprouted in cities and towns across North America. Some survived for decades, some for a couple of years; others never amounted to much at all. Now they have almost totally disappeared, and the machines they once fashioned of iron, bronze, and brightly polished brass are about as rare as copies of Hiscox' book.

Contemplating these early engines today, one could be forgiven for believing them primitive, even incomprehensible, in their stark, cast-iron simplicity. In fact, such engines represented a once-modern, practical refinement and expression of ideas and dreams that must have existed even before a Dutchman named Christian Huygens sketched an internal combustion engine to be fueled — perhaps "charged" would be a better term — with gunpowder in 1673.

The first practical internal combustion engine to be built and sold was constructed by a Belgian-born inventor who worked in Paris. There, in 1860, he applied for a patent on an invention that "consists first in the employment of illuminating gas in combination with air, ignited by electricity, as a motive force. Second, in the construction of a machine intended to employ this said gas." The inventor's name was Jean Joseph Etienne Lenoir.

In Lenoir's engine, the gas, according to the inventor himself, only served "the purpose of heating the air that is mixed with it. The air then dilated or expanded will act on the piston in the same manner as steam would in an ordinary steam engine." Lenoir's engine was, then, an atmospheric, not a compression engine. The pressure within the cylinder at the time of ignition was no greater than that without. It did not operate on the principles of intake, compression, ignition, and exhaust that characterized the more advanced four-cycle and two-cycle engines that superseded it and others like it. Even so, Lenoir's engine was recognized as the start of something momentous.

"The age of steam is ended," noted the *Scientific American* magazine. "Watt and Fulton will soon be forgotten."

In fact, Lenoir's engine looked like nothing so much as a steam engine. It was double acting, like a steam engine, and the fuel — equal parts of naphtha and turpentine — was admitted and exhausted by sliding valves, like a steam engine's. Ignition, however, was by high-tension coil and a pair of spark plugs. These Lenoirs turned some 80 rpm, and the first two that were sold were put to work driving machinery in factories. In the summer of 1861, a 1 hp Lenoir was installed in a 16½-foot boat in which the inventor took passengers for rides on the River Seine. Subsequently, Lenoir built a 2 hp boat and, in 1865, installed a 6 hp engine in a 40-footer. The latter, he noted, "was unsatisfactory because of the slow speed of the motor; we also used too much gasoline."

Fuel consumption was one of several problems confronting Lenoir. Another was excessive use of cooling water — a 3 hp Lenoir required a 350-gallon water reservoir — and rough running when placed under heavy load. As the engines' faults became apparent, some were broken up. Others were converted to become steam engines in reality as well as in appearance. By 1866, Lenoir sales had dwindled to almost nothing — it has been estimated that some 500 were sold — and the most fruitful work in Europe was being conducted near Cologne by Nikolaus August Otto.

Otto had built a Lenoir-type engine in 1861 and then embarked upon a methodical improvement program. This resulted in a more efficient, single-acting engine that used one-third less fuel than the Lenoir for the same output.

It was awarded a gold medal at the 1867 Paris Exposition. Working together with Otto on this project was an engineer and businessman named Eugen Langen. The two men formed N.A. Otto & Cie. in 1864, and, some eight years later, hired two brilliant engineers to help develop Otto's idea for what became the four-cycle compression engine as we know it today. Otto and Langen changed their company's name to Gasmotorenfabrik Deutz — Gas Motor Factory, Deutz — and concentrated on this project.

The two new engineers were Gottlieb Daimler and his self-effacing, lifelong friend, Wilhelm Maybach. In 1876, four years after they joined the *Fabrik*, a four-cycle engine was introduced. It compressed the air/gas mixture before ignition occurred and thus earned the name "compression engine." This device had been so refined by Maybach that it used less than half the fuel of a good noncompression engine, and it became a resounding sales success. The company quickly sold 5,000 of them — ½ to 3 hp — and also sold licenses for their production throughout Europe, England, and the United States. For a number of years, it was referred to as the Otto cycle engine, and Otto obtained a patent on the four-cycle principle that effectively stifled would-be competitors. In 1886, however, it became known that the whole four-cycle theory had already been described in an unpublicized patent by a little-known French engineer named Alphonse Beau de Rochas in 1861.

Here is the combustion process described by de Rochas: (1) induction during an outward stroke of piston; (2) compression during return stroke; (3) ignition at the dead point followed by expansion on third stroke; (4) discharge of burnt gases from cylinder during fourth and last stroke. When de Rochas' principle became known, Otto promptly lost his patent, some people began calling Otto engines de Rochas engines, and development activity was rekindled on a worldwide basis.

In 1882, Daimler and Maybach left Gasmotorenfabrik Deutz. They set up a tidy workshop in the garden house of Daimler's Cannstatt (Stuttgart) villa and designed and built a series of compact, high-speed — about 250 rpm — engines that surpassed anything in the world for power/weight ratio. Daimler wanted a small engine, light but still powerful enough for installation in vehicles of all sorts. On April 3, 1885, the Kaiserliches Patentamt — Kaiser's Patent Office — granted Daimler patent number 34926 for a single-cylinder, four-cycle gas engine. This was followed quickly by two more, including a 200-pound, 1.1 hp version. Stationary engines had weighed as much as 1,000 pounds per horsepower.

Although one of the singles was installed in a horseless carriage, Daimler wanted to continue development with a vehicle in which the engine's presence could be kept more secret. In October 1886, he was granted a patent for a motorboat. He and Maybach installed the engine in a lapstrake launch. They added some wire and a few insulators here and there, hoping to make people believe the vessel was electrically powered. They made their first test runs the following August.

Said a local newspaper: "Recently a boat has been circulating on the Neckar with eight persons aboard. It appears to be propelled by some unseen power up and downstream with great speed, causing astonishment on the part of bystanders."

Daimler, Maybach, and others — wearing three-piece suits and bowler hats — were photographed in this boat, and the photo was used in advertising the craft as early as 1891, when the company began selling motor launches of 18 to 35 feet, powered by engines of one to 10 horsepower. In 1890, Maybach designed a compact four-cylinder marine engine that developed 5 hp at 620 rpm and looked much like marine four-cycles in production two or three decades later.

While Otto and Daimler — and Carl Benz and Rudolf Diesel, too — were at work in Germany, inventors were also building engines in England and the United States, where George Brayton was selling his Brayton "Ready Motor" for use as a stationary engine. Another American maker of stationary engines claimed to have been the first to build a marine engine, or at least a marine engine equipped with electrical ignition, for Daimler's was of the hot-tube type. This American engine was an open-crankcase four-cycle built by the Union Gas Engineering Company of San Francisco. Union was affiliated with the old Philadelphia engine builder Globe, and claimed it built a motorboat in 1884, patenting a make-and-break igniter in June 1885. Globe, too, added a marine engine to its line. In 1896, the company's business doubled. This feat was repeated in 1897 and 1898, when Globe pointed out that no license was required to run a gas engine, as was the case for a steam engine. Nor were there any ashes to contend with.

Globe and Union were not alone in reaping a growing market. Ten years after Union ran that first motorboat, America's pioneering motoring magazine, *The Horseless Age*, estimated that there were more than 300 automobile companies in the United States and gave about as much coverage to their attempts to motorize boats as to their efforts to build automobiles. There was a substantial overlap in those days. It included such companies as Winton, Lozier, Stanley, Simplex, and even the Duesenberg brothers, whose marine-engine building prowess and subsequent liaison with the Loew-Victor marine engine company helped them finance their race-car program. Alanson Brush, renowned as the designer of the first Cadillac's single-cylinder engine, developed a marine version, too. David Dunbar Buick produced marine engines as well as bathtubs, before finally concentrating on automobiles.

Not all marine engine or automobile companies were attempting to build four-cycle engines. Before 1886, when Otto lost his patent, other inventors had already pursued other principles, in particular the two-cycle. Comparing the two engine types, it might be said that the four-cycle was really not as great a revolution as it seemed. It had much in common, two-cycle backers pointed out, with the steam engine, since it was clearly dependent on valves to admit and exhaust the mixture from the cylinder. Generally these valves were mechanically operated, but sometimes — especially before 1900 — the inlet valve was automatic, pulled open against a spring by the suction of the descending piston.

Two-cycle specialists believed their engines offered more efficient, simpler

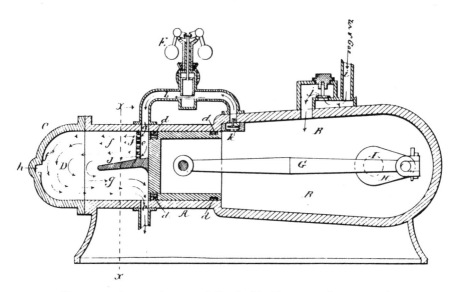

The patent drawing for one of Charles Nash's two-cycle engines, this one from 1888. "My improvements," wrote the inventor, "are to avoid the use of valves for the power-cylinder; to provide for admitting and exhausting the charge into and from the power-cylinder by the operation of the piston itself; to effect the displacement of the waste gases from the combustion chamber by the incoming charge, which acts to drive out the waste gases in a direction opposite to that in which the charge enters . . . ." It was men such as Nash who pioneered developments leading to the two-cycle marine engines that began appearing before the turn of the century. The engine pictured here, although inverted, is similar to a two-port one-lunger, having a check valve controlling entry of the mixture to the base. An additional check valve is used here at the transfer port. The piston is fitted with a deflector plate to direct incoming gases into the combustion chamber and away from the exhaust port.

operation. Often they referred to the two-cycle engine as a "valveless" to emphasize that it needed no complex, troublesome valve mechanism. They viewed the two-cycle principle as an ideal, a way virtually to double the output of the four-cycle, since the two-stroke fired at each complete turn of the crank, not every other turn. Many confidently believed the four-cycle would become entirely outmoded. The two-cycle's promise seemed similar to that offered by the double-acting steam engine when compared with the single-acting type, and some engineers zealously set out to make that analogy work.

Most of these two-cycle pioneers labored in obscurity and have been forgotten. Of them all, it is an Englishman, Sir Dugald Clerk, who remains most closely identified with the type's development. Clerk was a theoretician, patent agent, author (*The Gas and Oil Engine*), teacher, and internal combustion engine expert of considerable renown. His American counterparts, however, remain unknown. Most of them were amateur engineers, for the two-cycle idea seemed to attract more than its share of home mechanics. Lewis Hallock Nash is one of the few Americans whose name remains associated with two-cycle development. Working in New York City, Nash received some

40 patents relating to two-cycle design during the years between 1883 and 1897. Among these patents was a design for an engine that, in some respects, may be the direct forebear of the single-cylinder, port-controlled two-cycle that became popular during the first decades of the 20th century.

Nash's engine approached classical two-cycle practice by eliminating cam or eccentric operated valves. Its piston drew the air/gas mixture into the crankcase through a one-way check valve and delivered the mixture through a transfer port and through another one-way valve to the intake ports. In the cylinder, a deflector plate atop the piston directed the charge away from the exhaust ports and into the cylinder head. This was the same basic architecture that characterized the two-cycle marine engine built in America, Canada, and elsewhere for, in some cases, more than half a century. Such engines became so prevalent in this country that boosters of the type thought it a distinctly American invention. As for Nash, he built a line of two-cycle and four-cycle engines for the National Meter Company of New York. There is no evidence that any of these were used for anything but stationary purposes.

By 1910, however, the race to produce marine gas engines was well underway and several hundred companies were then in existence. Most built two-cycle machines, usually of one or two cylinders. Some offered triples or even sixes, but the mainstay was the one-lunger insofar as the two-stroke was concerned. Larger engines were mostly four-cycle, but debate about the virtues of each type was fierce, and each side had staunch disciples.

Summing up the arguments in *Marine Gas Engines* in 1910, Carl Clark wrote: "It may be stated as a general conclusion that for small, light engines where economy is not of great importance, and which receive little attention,

One of the assembling rooms, Sterling Engine Company, Buffalo, New York, 1911. These are four-cycle engines.

the two-cycle type is to be preferred. For single-cylinder engines the two-cycle type is decidedly to be preferred on account of the excessive vibration of the single-cylinder engine of the four-cycle type. For engines of large size, where fuel economy becomes of importance, together with increased reliability, the four-cycle type is probably preferable."

Sometimes, a company offered both types of engine and, when it did, felt the obvious necessity of pointing out the strong points of each. Erd Motors Corporation of Saginaw, Michigan, began manufacturing two-cycles in 1898, but by 1913, it was offering a four-cycle, too. Erd described its philosophy like this:

> Our two-cycle Motors are of the high and medium speed type, while our four-cycle Motors are of the heavy duty type.
> It has been known for years that the two-cycle Motor, properly designed and constructed, would give the very best of service for high and medium speed work . . . . In a four-cycle Motor of the same weight, bore and stroke, it would be necessary to have a number of small valves and weak parts which are subjected to considerable wear. This is a source of great trouble from the fact they cannot be made large and strong enough to withstand the work required of them.
> We strongly advocate the use of four-cycle Motors for heavy duty work, which we have for many years manufactured [sic]. In this type we are able to build the intricate parts large enough to withstand the hardest kind of usage and be subjected to the most severe wear from the fact that weight is not an essential feature, but rather an advantage, on account of the Motor being installed in boats of heavy construction, such as sea-going cruisers and commercial boats of all kinds.

The ideal represented by the two-cycle principle — that such an engine could develop twice the output of a comparable four-cycle — was never realized. Some engineers believed, however, that a 20 percent power advantage existed, and, quite possibly, they were correct. Bore and stroke dimensions were somewhat less for a two-cycle of a given power rating than for a four-cycle. Acadia's two-cycle, 5 hp make-and-break engine had a 4½-inch bore and 5-inch stroke. The Standard four-cycle, 5 hp make-and-break had a 5¾-inch bore and 6½-inch stroke. It also was heavier: 850 pounds versus 180. This is a dramatic difference, even when one considers that the Standard's weight includes a reverse gear weighing 100 pounds or so. The two-cycle engine was not frequently fitted with these devices, being able to run just as well backward as forward. "Reversibility" was another advantage claimed by two-cycle engine makers, and most owners quickly learned how to "reverse on the switch." The tactic was not 100 percent sure, as we'll see in a later chapter, but this manipulation of the ignition switch worked most of the time. When it did not work, tragedy could occur. There are still those on Cape Cod who will tell you that, each year in the old days, a crewman aboard one of the big cat-boats would be killed, crushed between mast and dock when the engine failed to reverse as planned.

The two-cycle one-lunger developed about 1 hp per 36 pounds of weight, sometimes more and quite often less. The Erd 5 hp weighed 150 pounds, the Kennebec weighed 297, the Ferro 210. It is difficult to compare the weights of

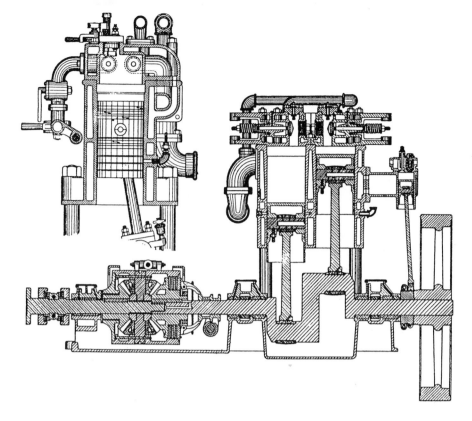

Section through a Standard two-cylinder four-cycle marine engine, circa 1910. The overhead valves were mounted horizontally and the make-and-break igniter was located centrally in the head, directly above the piston. Said Standard: "This feature . . . brings the distance of point of ignition to the various points in the combustion chamber much closer than in the T-head engine, and in the L-head type the spread of flame through the gas must pass from the point of ignition into the cylinder and around there into the combustion chamber over the exhaust valve." Note the long piston skirts with four rings, the relatively large piston surface designed to increase the piston's life. Liberal oil grooves aid lubrication. The crank and rods were made of steel billet to Lloyd's specifications. The rod bearings were beefy bronze castings fitted with metal shims so wear could be compensated for easily. The drawing shows a vaporizer suitable for gasoline, benzine, or No. 1 distillate, but those who preferred a float-feed carburetor could order a Schebler. Like other four-cycle marine engines, this Standard is fitted with a reverse gear. The two-cylinder Standard was offered in 8, 10, 12, 16, 20 and 30 hp models. The 8 hp weighed 1,000 pounds; the 30 hp weighed 4,190 pounds.

various makes precisely, since some manufacturers included a muffler, shaft couplings, and other accessories in their figures, while others did not. Certainly, the power/weight ratio was acceptable for marine use.

Few builders of two-cycle engines were concerned, as was the Elbridge Engine Company of Rochester, New York, with reducing their engines' weight. Elbridge, however, catered to a market for speedy pleasure craft and racing boats. Instead of iron pistons and cylinders, the Elbridge used aluminum/copper alloy. There was even a "featherweight" series that included a four-cylinder, 40 hp model, which, stripped of accessories, weighed only 150 pounds. It was suitable, in fact, for installation in airplanes.

The level of technology attained by the one-lunger is, in a sense, illustrated by a brief comparison with aircraft engines built during World War I. By war's end, airplane engines were developing 2.5 hp per pound or even slightly better. To obtain light weight, parts of the engines were made of such metals as chrome vanadium steel — light, strong, and expensive. It was ideal for connecting rods. Cylinders were machined to extreme tolerances to save weight and had water jackets welded on, a weight-saving process originated by Mercedes before World War I. No such weight-reducing measures, or costly steels, were necessary or particularly desirable for the one-lunger. Rather, it was common practice to add weight to castings or other parts as a means of ensuring more than adequate strength. Unlike airplane engines, one-lungers went a long time between overhauls.

The Fay & Bowen Company of Geneva, New York, summed up the situation neatly in its 1908 catalog: "In building our motor, we have not tried to see how light we could make it. It is not a heavy machine, but rather than make it light at the expense of stability, we have preferred to put enough iron in its frame to insure rigidity and safety, and withstand the continuous explosions." The company did not bother to include engine weight in its specifications tables.

Although four-cycle engines rather quickly came to dominate the automotive field, the two-stroke gained a secure niche for use in small or medium-sized boats, especially workboats. This preference for two-cycles prevailed everywhere but in California and the Northwest, where, beginning with Union, four-cycle engines predominated among both fishermen and yachtsmen. A number of manufacturers of fuel/oil engines with hot-tube ignition also existed, especially in the Pacific Northwest.

Considering the possible maladies of a four-cycle marine engine in 1905, *Powerboat News* described what an owner might face if his engine refused to start:

> The inlet valve may have become stuck. You remedy that, and it still refuses to go. Spark is good, carburetor is correctly set, and everything apparently in good order. After tiring yourself out trying to locate the source of trouble, you might possibly find the trouble yourself, but chances are you would secure the assistance of someone familiar with the make of engine, who might discover the exhaust valve stem stuck where it passes through the guide or bushing, and, after loosening and oiling with kerosene, the trouble might be righted.
>
> Again, the valve might not seat properly, owing to carbon deposits on the valve seat, or you might find the valve worn and requiring regrinding to hold compression.
>
> All these details, including mixture and ignition troubles, are adjuncts to the 4-stroke engine, not counting on the liability of cams and gears slipping.

For the man who used a two-cycle engine, such considerations as these would have been unthinkable. He operated in a world where low operating costs and, above all else, self-sufficiency were the most important factors relating to his boat's engine. In such a world, the two-cycle one-lunger was

Campbell four-cycle 5 hp motor with side plates removed, jump spark (5 hp at 500 rpm, 5 x 5½, weight 375 pounds). Circa 1920. Campbell Motor Company, Wayzata (Lake Minnetonka), Minnesota.

uniquely well suited. In some respects, as we shall see, it has not been surpassed even today.

Compared with the four-cycle, the two-cycle marine engine had one overwhelming virtue. It was simplicity. Valve gear was dispensed with, replaced by intake and exhaust ports within the cylinder that were alternately covered and uncovered as the piston moved up and down. This simplicity was, however, *relative*. It did not mean that the two-cycle was troublefree, only that such troubles as it was heir to were usually quickly repairable, even in an open boat at sea. It was commonly said of the two-cycle that it could be repaired with a piece of baling wire and pliers. In Nova Scotia, they replaced those items with a "cod napper and a peavey." The former was a club used to subdue a flapping codfish, the latter a lever used by lumberjacks to move logs.

Bob Merriam is an experienced boatman who once owned a small schooner equipped with a two-cycle, two-cylinder engine built by Acadia in Bridgewater, Nova Scotia. He used this boat every day, winter and summer, for three years. "I don't think you could honestly ever get more than an hour's running without changing something," he remembers of that particular engine. "But it never was anything serious."

Like the old fishermen he had met, Merriam kept a mayonnaise jar aboard his boat. It contained key spare parts bathed in oil so they would not rust. Whenever he needed a part, he simply took one out of his mayonnaise jar, put it on, and continued. He did not need a factory repair manual, special tools, or diagnostic equipment, which he would have needed with a modern engine. He did not have to worry about the ticklish business of removing a broken valve spring, as he might have done with a four-cycle engine of an earlier era.

If simplicity and self-sufficiency characterized those who bought and used two-cycle engines, they were also traits of those who manufactured them. Not all the companies were large, and quite a few hardly deserved the name "manufacturer." The one-lung marine engine was as likely to be built by a little machine shop or foundry as by a large factory. Most companies were housed in small buildings, and some had foundries no larger than a living room. The yard outside was often cluttered with rusting iron and debris. Inside, many of the tools were homemade.

"You would be amazed," said Bob Merriam, who once visited a number of tiny Canadian engine builders, "at the contrivances those guys put together to create these engines."

Some of the shops were so small that they contracted substantial amounts of work to outside suppliers. The Howard Motor Company of Philadelphia, for example, had its connecting rods made by William Cramp & Sons, the same yard that supplied many of the Navy's battleships. It was not unusual for a medium-sized company — one that built 2,000 to 3,000 engines per year — to order crankshaft forgings from a specialist in this work and then finish the grinding and polishing in-house. Almost everybody ordered carburetors and ignition coils from those who specialized in such esoteric components.

The men who ran the engine companies and who worked in the shops grew up in post-Civil War America, a time when the country was beginning an awesome shift from a largely rural society to an increasingly urbanized and, perhaps more important, mechanized way of life. Young men who found themselves more fascinated with machines and steam engines than with farm work were able to make places for themselves in a changing world. For the most part, those who designed the first marine engines — like those who designed the first automobiles — were born mechanics, tinkerers of real talent rather than trained engineers. "I have no idea how he did it," said Robert Scripps of his father, William. William Edmund Scripps was the son of the founder of the *Detroit News,* and eventually he became president of the newspaper himself. He also founded a company to build Scripps marine engines, four-cycle machines of the very highest quality. William Scripps never attended college and had no knowledge of mathematics. He designed his engines from scratch and made the patterns. He could do everything required except make castings, a common problem for most pioneers, who hired skilled foundrymen. Scripps's company built the engine for the *Detroit.* The motor-

Torrey Roller Bushing Works, Bath, Maine, circa 1925. (Courtesy Maine Maritime Museum)

boat completed the first transatlantic crossing by a gasoline-driven vessel in the summer of 1912.

It is doubtless true that for every dreamer who succeeded in building a viable marine engine, many more failed. Sometimes, those who evolved a workable engine started their own company to build it. More often, perhaps, they became associated with machine shops or foundries of long standing. That was the case with Frank P. Lord. He was a tinkerer who designed an engine for Haviland Torrey, whose father had begun the Torrey Roller Bushing Works in Bath, Maine, in 1869. Lord designed an engine, using a

Kennebec engine, Maine Maritime Museum.

Palmer as a model, and that was how Torrey got into the business, adding a one-lunger to its other products. The Torrey shop was located on the shore of one of Maine's great rivers, so the engine was named the Kennebec. Torrey built Kennebecs until World War I, at which time the forge that supplied crankshafts became too busy with war work to continue doing so.

There were trained engineers who worked on marine engines, but their names were, for the most part, ignored by those who employed them. When W.T. Ritcey began Acadia Gas Engines, his first catalog of 1907 had this to say about the engine's designer: "The designs of our engines are original and designed by our superintendent who for many years was connected with one of the first gas engine manufacturers in the United States and who for the past four years superintended construction and building of a well-known engine manufacturer in Maine . . . ."

It is likely that such trained and experienced engineers were employed by a number of different companies, just as was the case with early automobile engineers, most of whom knew each other and pursued careers first with one company, then another. Stanley R. DuBrie was one such man. A peripatetic mechanical engineer who graduated from the University of Ohio in 1874 and worked in the Detroit area, he created a successful hot-tube marine engine well before 1900. Later he designed the Sandusky, a two-cycle; the DuBriemobile, a four-cycle built in Ontario; the De Loach and DuBrie-Caille automobile engines, both two-cycles; the Bense, the Waterloo, and the DuBrie. He also had a part in creating the Gray, introduced in 1906, before going on to design the Peerless marine engine. "Perhaps no engineer," said Peerless of their chief engineer, "has had such a line of experience with such noted success."

Not all two-cycle engines were successful. Although operation depended to a great extent upon the owner's skill, most brands acquired reputations for either reliability or cussedness. Around Camden, Maine, it was said that a Knox engine — built by the Camden Anchor-Rockland Machine Company — would keep running almost submerged. But fishermen around Boothbay occasionally used to say of the Boothbay engine that, "If you look up and see one coming at you from out of the sky, don't worry. It'll stop before it gets to you."

For the most part, fishermen followed the lead of the first man in their area when it came to choosing an engine. If the example of a given type seemed to operate well, that was generally enough for the rest of the men. Likewise, news of a troublesome machine spread faster than gossip, and a stigma would remain forever attached to that brand. Gerald Smith, a Marblehead (Massachusetts) fisherman, remembered what happened after an Italian skipper in Boston had trouble with a Mianus engine.

"After word got out," said Smith, "you couldn't sell a Mianus to an Italian in Boston. It had to be a Lathrop or maybe a Bridgeport, which they pronounced 'BreegeAport,' and that was all there was to it. You couldn't change them, either."

Those in Smith's area who tried to change the fishermen's minds were the salesmen from Boston's Commercial Street. They journeyed about Boston, the

North and South Shores, and did their best to convince people that their particular machine was best. Those who heard them recall that they were smooth talking and quick witted. "Why," said one, "those boys could have taught an automobile salesman a thing or two." Real sales skill was often important to the drummers. About the time of World War I, a good 5 hp one-lunger cost some $185, not cheap in a day when a boatbuilder or fisherman might earn $50 per month.

During the first years of the new century, the move to power was, like some massive cast-iron flywheel, gaining momentum. Aboard coasting schooners, whose skippers were always looking for ways to save money, the gas engine permitted substantial labor savings. "The greatest single aid to short-handed operation," wrote John F. Leavitt in *Wake of the Coasters*, "was unquestionably the gasoline hoisting engine. This was usually installed in the forward deckhouse, if there was one, and, if not, under a box or cover of some kind adjacent to the foremast. The favorite was the Fairbanks Morse 'Bulldog,' a one-cylinder, two-cycle machine which was as powerful as it was noisy. These engines were hooked up in various ways to windlass and pumps, and also operated two winch heads at the ends of a horizontal shaft on which the halyards were wound when hoisting sail or lifting out cargo." The schooners' yawlboats were also fitted with one-lungers able to push them at 3 or 4 knots when called upon.

To the average coastwise fisherman, the promise of the one-lunger was irresistible. It meant an end to rowing heavily laden skiffs and dories, a way to get into port when the winds fell calm. It meant an easier life. Rigs began to be reduced in size as the necessity of taking advantage of every wisp and zephyr disappeared. Work launches and other types of boats, intended solely for

power, began to appear. Traditional American sailing craft began to be modified so an engine could be installed.

Charles Sayle, a longtime Nantucket resident, remembers that the first engine came to the island around 1902 or 1903 and was mounted in one of the catboats used there for scalloping. Sayle installed a 1½ hp engine in his own dory. It wasn't long before many of the island's fishing boats had been converted to power.

"I can still hear the one-lungers that first day of November when the scalloping started," Sayle said. "At the break of dawn, the whole town would wake up to them. There'd be 40 or 50 of 'em cracking away all over the harbor."

Thus did the great shift from working sail begin. In the New England catboat fleets, the boats were modified as necessary to take the engine bed, fuel tank, and shaft log. Sometimes this meant cutting out half the boat's centerboard, but this, and its attendant loss of windward ability, was counted a small price to pay. It wasn't long before most boats were converted, generally with an 8 hp one-lunger. This size of engine and a 22-inch propeller were judged about right for a 23- or 24-foot catboat. Often, after installation of an engine, the mast was removed for winter scalloping and replaced by a stub used only for mooring.

Having power in a boat required the skippers to learn some new work methods. Under sail, all the vessels tending nets or drags tended to drift in the same direction. Powered boats could move about at will, crisscrossing and fouling their neighbors' gear unless care was exercised.

Simple as the two-cycle engine was, it usually was the most complex piece of equipment its owners had ever seen, and it operated on principles with which nobody was particularly familiar. Fishermen have always been self-reliant sorts, however, and the two-cycle one-lunger soon became known for the ease with which it could be run and repaired. These were the timeless virtues of the type. Even at sea, a one-lunger could be brought back to life. Joseph Chase Allen grew up among the fishermen on Martha's Vineyard and remembered what happened one breezy day when he went fishing with one whose Lathrop backfired and blew out its head gasket in small pieces.

> After that, of course, it wouldn't run. We made our own gaskets in those days out of asbestos paper, which was about as thick as thin cardboard and made out of material of about the same consistency. The way we did it, we took a sheet somewhat larger than 8½" x 11", took the cylinder head off and laid the sheet of paper on there, and then, with a hammer, we'd tap around the sides and the bolt holes and so on, until we could take it apart and it would fit. We had some of that paper aboard. We had the tools. But here was the boat standing on her head one minute and on her tail the next and then rolling from side to side and doing all sorts of funny things. The man and I laid down flat on the deck, took that cylinder head off, made a new cylinder head gasket, and put it back again, all the while barking our hands as the wrenches slipped on the nuts and chasing the things that slid into the scuppers every time the boat rolled.

The two-cycle's repairability was not to be taken lightly. It lay at the core of the one-lunger's appeal, and nobody expected to have to call up his engine dealer in order to get something fixed. With each step away from the basic two-cycle engine, with each new technical advance, engines became less approachable. You can't fix a modern outboard, with its electronic components, or a fuel-injected diesel, with baling wire and pliers. The old one-lunger was a shipmate a man could handle on his own, and it represented an independent lifestyle that has almost disappeared today, when each member of society must depend upon someone else. When a modern motor's ignition module fails, one needs to call the dealer, who must check his computerized parts list and, perhaps, make an appointment with a factory-trained technician for the engine's repair. Because the one-lung, two-cycle engine lay at the opposite extreme from all this, it was still being installed in boats late into the 1920s. A few companies even continued building two-cycles up to, and after, World War II.

One of these companies was Acadia Gas Engines, which regularly built its line of engines until the 1960s. Many remain in daily use even now. In the winter of 1980, the company received a call from a trapper who ranged well beyond the Arctic Circle in an Acadia-powered boat. He called to ask if he could purchase a new 10 hp engine and was bitterly disappointed to learn the answer was no.

"He's way up there, isolated," said Eric Whynacht, the firm's sales manager, "and his life depends on an engine that he can keep going himself. With his old two-cycle, he's not afraid to go out a couple of hundred miles with nobody

The Acadia plant in June 1981. Peaked roof section is original building.

within earshot of him." But there are not many men left like that trapper, and there was little that Acadia could do to help him.

The great decline of the marine two-cycle engine began with the ever-increasing durability of four-cycle types as engineers perfected valve mechanisms. Expanding markets and a stronger emphasis upon speedy delivery of fish to shore also caused a move toward higher-revving four-cycle machines. Strangely enough, the reversing gear of the four-cycle was also a factor in the two-stroke's demise. Fishermen were not particularly interested in backing up their boats, but they were most interested in rigging the reverse gear so that it could also be used to haul lobster pots or trawls.

Some old-timers recall that it was the Depression that sealed the two-cycle's fate. Nobody had any money to buy a new one-lunger, and, instead, people found discarded automobile engines that could be converted cheaply for use in a boat. Parts for the engines could be had equally cheaply, sometimes for nothing. Gradually the old two-cycles were abandoned, sunk for use as moorings, or left to rot, paint flaking and water jackets splitting, in a field behind the house. "After World War II," said Charlie Sayle, "the outboard just took over."

By the time that happened, the two-cycle inboards had been gone for over a decade and most of the companies building them had long since ceased to exist. Today, only the Lunenburg Foundry is left. The engines themselves, with their tall cylinders, chunky bases, and brass parts, look like what they are, relics of a vanished age, mementos of a lifestyle one must struggle to find in America anymore. It was only 50 or 60 years ago that the old one-lungers could be seen and heard in many places, thumping out their throaty tunes under the sure hands of thousands of independent fishermen. These days, there is almost nobody left who even remembers.

All over the world skilled engineers and designers are giving the best years of their lives to still further perfect the gasoline engine and it can be predicted with certainty that the fruit of their labors in this direction will be clearly shown in still greater improvements and developments as time goes on.

Marine Engines and Equipment, A Practical Treatise on Correct Design & Construction for Layman, Experienced Boatman & Boat Builder,
The Ferro Machine & Foundry Co., Cleveland, 1912

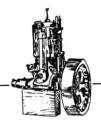

2

Portraits

OF THE HUNDREDS, even thousands, of companies that once built marine gas engines, relatively little is known. It is almost as if those who pioneered the industry stepped out of the woods one day, created something where, before, there had been nothing, and then disappeared. What they created were engines, gasoline engines, machines that inhaled air with breathlike gulps and regularity, mixing the oxygen with fuel and converting the result into heat and energy in a manner not entirely unlike the functioning of a live creature. Before 1885, few men had even imagined such things. Ten or 15 years later, men all over the world were staking their futures on building them.

If today they seem to have emerged from nowhere, this is because so few of them ever bothered to document their experiences. "Oh, the old-timers never would have dreamed," said the son of one early engine maker, "that anyone would ever want to know about them."

Some did keep records and some even wrote down the stories of their lives in little diaries, sketches full of wonder at being part of a movement that so obviously was affecting virtually everything. But most of this material eventually was lost. Often it was destroyed in the fires that forever seemed to plague the foundries and shops. Much of the literature was simply thrown away, discarded in that dead time between when an object has become a useless, outmoded relic and when it becomes an "antique," possessing both nostalgia and increasing monetary value. It has taken a generation or two for old gas engines to catch on.

Many of those who became gas engine builders had worked in machine shops or forges, in bicycle factories or steam engine shops — anyplace at all where a man might develop a sense of iron, steel, and some conception of the internal combustion process. Some, as we have seen, were engineers. Yet, for every DuBrie or Brush or other man with formal book learning, there were dozens who seemed to understand, as if by some divine osmosis, the principles of gas flow, mechanics, and electricity. So it was that, from little shops in steamy Louisiana bayous, from up-to-date factories in cities by the Great Lakes, from boatyards around Boston and machine works in San Francisco and along the Maine coast, marine gas engines emerged into the world.

Following are portraits of six marine engine companies and the men who created them.

PALMER BROTHERS

The first two-cycle marine engine on the East Coast is generally thought to have been built in 1894 by two brothers in Cos Cob, Connecticut. Their names were Frank and Ralph (Ray) Palmer. Throughout the brothers' lifelong partnership, Ray Palmer — born in 1860 — was responsible for invention, design work, and shop supervision. Frank Palmer handled the business details and promotion and attended the boat shows his younger brother shunned. Frank was older than his brother by three years, and was sometimes referred to as "the executive head of the house."

Those who knew the Palmer family attributed much of the brothers' interest and success to their father, Rex, who was known in the Greenwich area as owner of the largest cider mill in Connecticut. Cider presses were among the products once manufactured by the Palmers, and both brothers may have become familiar with these powerful, hydraulic devices at an early age. For years, the same Mack truck that used to deliver Palmer engines also carried barrels of fresh cider.

In 1888, Frank and Ray Palmer established their own business. They set up shop at Dumpling Pond in North Mianus, Connecticut. There they assembled bicycles and manufactured telephones and electrical equipment. In their spare time, they continued to exercise an interest in rowing boats, and it was this latter hobby, the brothers told a local newspaper in 1909, that led them to experiment with gas engines. Apparently, these were serious experiments, not after-hours tinkering, for the brothers estimated they spent $5,000 on their research.

It is unlikely that during the motor's development Ray Palmer devoted much time to telephones or anything else in the Dumpling Pond shop. He designed and constructed a 3½" x 3½", two-port, two-cycle engine with a primitive, wick-carburetor gas producer. This was simply a metal tank full of gasoline and cotton waste. The material soaked up the gasoline and gave off the resultant vapor into the engine through a one-inch-diameter pipe covered with wire gauze to prevent backfires into the gas producer itself.

The prototype engine was redesigned at least once or twice. Throughout its development, Ray Palmer was determined to create a foolproof sparker and ignition system, for this component, he knew, had long retarded gas-engine development. At a time when most igniters were of the long-contact, slow-release type, Palmer designed one that broke contact with a quick snap, giving a fine, hot spark. In 1896, he experimented with a jump-spark system, using a spark plug made in France. The plug lasted an hour before its porcelain cracked, but Ray Palmer believed jump-spark would eventually replace make-and-break once better plugs were developed.

He continued to refine his make-and-break system, and this work led to what the local paper called "the first motor of its kind ever built in New England and one which with small improvement found sale more readily than it could be prepared for market." Its impact on other would-be manufacturers was profound. Many were influenced by the Palmer and many probably copied its essential features.

The first Palmer production model was built in 1895. The brothers installed one of these 1½ hp engines in a specially built 15-foot Whitehall and used the boat around their dock area for 15 years. The motor survives to this day, owned by a former Palmer salesman. The 1½ was followed in 1896 by a 3 hp. These engines were put together, at the rate of three or four per week, in a small factory operated by water power from the Mianus Falls. In those days, the brothers bought some components from outside suppliers. Crankshafts, for instance, came from the Cape Ann Forging Company, which retained many Palmer dies during the years before large-scale production was begun. All the electrical equipment, however, was made in-house initially. Palmer wound its own ignition coils for the first five years of its existence.

Much of the Palmers' early success was due to the favor their engines found with local oystermen. The first Palmers sold were used in workboats, and for some years the company bought boats and installed engines right at the factory, for it had become apparent that this was one way to sell more motors. A decade after production began, the brothers estimated that fully half of their enormous output — some 10,000 engines by 1905 — had gone into workboats. The Palmer engine's blue-gray color became familiar in harbors all over the country and, eventually, all over the world.

To keep pace with demand, the Palmers enlarged their factory three or four times until, in 1901, they purchased four acres in Cos Cob for a new facility with ample dockage on the Mianus River. Here were built a large machine shop, boatshop, shipping room, foundry, and assembly room. All were outfitted with the very latest machine tools. From the new factories flowed an increasing number and variety of engines, both two- and four-cycle. By 1916, some 30 models, ranging from 2 to 75 hp, were available, and it was estimated

Palmer two-cycle, circa 1910.

that 50,000 Palmers were then in operation. By 1924, 75,000 of the "Fisher-man's Friends" had been built. The brothers became wealthy and invested heavily in local real estate, especially rental housing. It was said that one could always tell a Palmer-owned house. The brothers bought their roofing material in large quantities, so all their houses had roofs of the same color.

Ray Bolling joined Palmer Brothers in 1933, not long after his graduation from Bowdoin College in Maine. He had grown up in the Greenwich area among Palmer workmen and fishermen who used Palmer engines. He didn't want to pursue the Wall Street career outlined by his family and quit his first job in New York two weeks after he started. He walked up to Frank Palmer at a boat show in the Grand Central Palace Hotel and said: "Mr. Palmer, I want to get a job with you."

Palmer remembered Ray Bolling as a boy with an intense curiosity about the company's products. "Well, sonny," he told him regretfully, "we're just not hiring anybody. Things are bad."

Bolling said he would work for nothing.

"Come in Monday," Frank Palmer answered.

In fact, for insurance reasons, they had to pay Ray Bolling a starting wage of 25 cents per hour. He was put to work in the assembly room where, under the guidance of an old Irish engine fitter named Brennan, he learned how to build engines. They were four-cycles by then, but the methods used were the same as for the earlier two-strokes. The finishing crews received a base with fitted crankshaft. They went into the parts room and removed all the other pieces necessary to complete the assembly. They worked an eight-hour day during the Depression, six days per week. Each team built one or two engines a day.

After he learned to build engines, Ray Bolling was sent to work in the machine shop, and eventually he became familiar with most aspects of the company. When World War II began, he was responsible for putting the 30,000 parts in the parts room onto a Remington Cardex System, the first time they had ever been cataloged. During the war, Palmer Brothers built its Model LH, a four-cylinder for installation in ships' lifeboats. These sturdy little engines were turned out at the rate of seven a day.

Frank and Ray Palmer were cast in the traditional mold of the hard-working, cost-conscious, independent men who pioneered the gas engine industry. By the time of Frank's death in 1944, the company's direction had already begun to shift. Business continued, but it was never quite the same again, despite family involvement. After the war, Palmer Brothers was sold to a New Jersey company named Columbia Air Products. Ray Bolling remembers the purchase as one made by Columbia primarily for tax purposes. Within a year and a half, the company was $1.5 million in debt. Needlessly made castings were stacked in the yard under a mantle of dirty snow. By the time he died in 1953, Ray Palmer was probably much saddened by what had become of the company.

Bolling and a number of associates managed to purchase Palmer Brothers at a bankruptcy auction, narrowly outbidding the junk dealers who hoped to get the machinery and iron. The new owners hired back 25 old-timers and began an intensely difficult rebuilding process. Their chief product became a rugged engine based on an International truck block. "It made a beautiful engine," Bolling remembers.

Bolling ended his association with the company — then called Palmer International — after an unsuccessful attempt to sell out to International Harvester in 1968. The business was continued for a few years by his former partner before ceasing operation. Of the company's great days, almost nothing has survived. All the drawings and records were destroyed when they seemed to be of no further value to anybody.

Bolling himself had restored the first 1895 Palmer made by the firm, and this engine, painted white and labeled "the granddaddy of them all," was for years exhibited at boat shows. Now, even that engine has disappeared, but Bolling says he hopes he can find it again and give it to a museum.

HAWBOLDT GAS ENGINES

Frank Hawboldt was born in 1876 in Marys Cove, Nova Scotia, a tiny village some two miles from Chester. He was the son of a merchant who owned a pair of lumber schooners, which sailed regularly for Halifax, trading lumber for store goods. Frank Hawboldt had relatively little interest in carrying on this business. He was handy with watches and determined to become a jeweler, so he married a girl from Chester and set up his jewelry shop there.

Hawboldt maintained this business until 1900, when he became fascinated by the first gasoline engines he saw in town. Some came from the United States. Another, the Toronto Junction, had been built in Canada. "My father looked at these," said Frank Hawboldt, Jr., "and decided to build one himself."

Hawboldt did just that, designing an engine from scratch. He made his drawings on brown wrapping paper and proceeded to build all the necessary wooden patterns. Then he began work on the parts themselves out in his barn, using a couple of old lathes. This engine was a single-cylinder, jump-spark two-cycle, and it apparently worked well enough for Hawboldt to give up his jewelry business and plan a future around the manufacture of marine gas engines. He needed professional guidance for only one aspect of production, and this was provided by a Scotsman named John Douglas. Douglas had a comprehensive knowledge of foundry practice, and he helped Frank Hawboldt set up a foundry in the barn. There the first production Hawboldt was built, a 4 hp one-lunger that was still around, in parts at least, a half-century later.

"We don't know how many engines were built in the barn," says Hawboldt's son. "He moved to a new shop where he had water power to run his equipment. Previously, he'd had a stationary engine to power his tools, but he figured he had about 50 hp of water power available at the new site. He used that to his advantage and gradually built up his business."

The design of the engines remained largely unchanged as the years passed. It was not until the 1930s that Hawboldt began adding nickel to harden the engine's iron cylinders. By then, other companies had done so for years. Some, however, never did. When Frank Hawboldt, Jr., graduated from Nova Scotia Tech as a mechanical engineer in 1932, he went to work for his father and attempted to make some improvements on the engine. "We did a little bit," he recalls, "but there wasn't too much you could do to it. You had to keep it simple. We did improve the sparking gear and points and began using stainless steel for some parts that had been ordinary steel." The younger Hawboldt tried a few other things, too, as we'll see in the chapters immediately following, but he learned that the old design and materials just didn't permit, or need, much change.

Like most Canadian engine builders, the Hawboldts found that their

A two-cylinder Hawboldt engine. (Courtesy Chris Hawes, Heritage Foundation)

business was primarily local, although a certain number of engines were sold in that seemingly inexhaustible market for the one-lunger — Newfoundland. The 4 hp cost $95 in the early years, but the elder Hawboldt liked to tell a story about the time an old fisherman, unfamiliar with handling cash, went to Chester from Indian Point to buy one.

"Somebody told him that he shouldn't pay more than $100 for the engine," says Hawboldt, Jr., "and that was the only number he had in his mind. He asked Father how much the engine would cost, and Father told him $95. The fisherman said: 'No way I'm gonna give you $95. I've got $100 with me, and that's all I'm gonna give you.' "

The Hawboldt company began with one employee, but by World War I, about 25 people worked there. During those years, and up until the Depression, the two-cycle engine was entirely adequate for the inshore fishing then conducted. Hawboldt, Jr., remembers there was a change in the fishing pattern that began in the late 1920s or early 1930s. Then fishermen began going farther offshore, and the two-cycle, even a two- or three-cylinder, wasn't fast enough for that new sort of work.

"They had to have more power," says Hawboldt, "and the marine engines of that time would have cost too much. They started converting car engines then, and they did the trick just as well."

A three-cylinder Hawboldt engine. (Courtesy Chris Hawes, Heritage Foundation)

With the two-cycle's decline, the Hawboldts expanded a line of hardware and other castings. Engines were made on occasion until 1954, however. Most, if not all of them, went to Newfoundland.

Frank Hawboldt, Jr., ran the company after his father's death, and he remained in Chester after he sold the works and retired. The firm still exists, manufacturing hardware, on the site that Frank Hawboldt selected years ago. Musing about the reasons for his father's success in this modest engine-building endeavor, Hawboldt, Jr., smiled and shook his head. "There's no explaining it, is there?" he asked. "Except to say he was a mechanical genius."

THE BOOTHBAY ENGINE COMPANY

Of his grandfather, Henry, Bob Rice remembers, "there was nothing he couldn't do if he wanted to."

Henry Rice was born in 1874 and grew up by the boatyard in East Boothbay, Maine, where his father, a cabinetmaker, turned out about a dozen small sailing craft. As a child, Henry also frequented the shipyard owned by his uncle, Norman Fuller, who created some of the hundreds of wooden vessels built in Maine. "I got my feet used to those yellow chips early," Henry Rice told a reporter once, "and I would stand by the hour and watch whose whipsaw men, one on top and one below, sawing out the timbers." Once sawed, the timbers were drawn to the appropriate spot by oxen. "Why, in those days," said Henry Rice, "even the oxen had to get a certain amount of education, and a shipyard was nowhere unless it had oxen to drag the timbers around."

Wood, saws, oxen, and ships were second nature to Henry Rice, as they were to his brothers Frank and Will. The three established Rice Brothers in 1892 and, during the subsequent decade, turned out catboats, sloops, and schooners. Many of these boats were fitted with inboard engines, and one day Henry Rice decided he was tired of delays in engine delivery. He was a man who liked to deliver a completed vessel exactly when he said he would. In March 1902, he decided it was time for the company to begin making its own engines.

Using intuition and his experience with other makes of marine engines, Henry Rice designed the Boothbay engine himself. His brothers were not enthusiastic about the project, but Henry was the senior of the trio and had the last word. Will secretly harbored a desire for a fast motorboat and never objected very vocally. The first engines were built in the spring of 1903, two-cycle, 2 hp machines with make-and-break igniters. A 6 hp followed, and then a twin-cylinder 12 hp. In 1911, a 60 hp, four-cycle type was added, and there were other four-cycles of smaller size as well.

Rice's most pressing problem in starting production was recruiting a workforce. "Few men in this area," said Bob Rice, "knew steel." Initially, the engine works at Rice Brothers employed only five people, and Henry Rice considered himself fortunate to have them. From Back Narrows came Reginald Matthews; from East Boothbay, Charles Spear. They knew steel. They were first-class machinists, perhaps the best in the region, even including the great iron works at Bath, where everyone knew steel. Henry Rice hired Matthews and Spear at the start of his engine-building venture, and they remained with the company for 30 years. Nobody knew how they acquired their unusual metalworking talents.

"I often heard from my own father," Bob Rice said, "that it just came natural to Reggie, that it was something he was born with. He could take a piece of steel and do anything you wanted done with it." Charles Spear's skill was perhaps easier to explain, since he had worked in shipyards building steel vessels.

Boothbay Engine Company was a subsidiary of the Rice Brothers shipyard. Initially the engine business was conducted on a small scale, with most of the engines going into launches and small yachts. Between 1902 and 1911, it is unlikely that more than 15 engines were built each year. By 1911, the Rice Brothers' reputation had spread well beyond Maine, and business increased substantially. That year, a new foundry and assembly room were built, and the staff increased to 25. One hundred engines were built in 1911, and thereafter, production equaled or surpassed that pace.

All of this activity, both in the engine works and in the yard itself, was carefully documented by Henry Rice. He was, among other things, a skilled photographer. Each morning before he began work, he went about the fac-

Rice Brothers Company plant after new construction. One story was added to main shop (background); brick building is machine shop; engine shop and foundry are at right of machine shop. September 1916. (Courtesy Bernard M. Rice)

tory, recording each new ship laid down, progress on others — the yard was the first in Maine to build a steel ship — and in the engine-building shops. Nearly all of these early photographs were destroyed by fire in 1917. The fire did some $450,000 damage and, to recoup some of the painful losses, only partly covered by insurance, Henry Rice sold the Boothbay Engine Company. The equipment was purchased by a local machinist named Thurston, who continued production, in very modest volume, until 1924. It is said that all the engines built by Boothbay were painted red and that Thurston's were painted gray. Bob Rice says, however, that, as a child, he also saw some gray Boothbay engines.

After the sale of the Boothbay Engine Company, Henry Rice continued his shipbuilding with his brothers. He followed a schedule as regular as the tide, working each day from 6 a.m. until 6 p.m. Bob Rice recalls that even when his grandfather was 84 years old, he was a hard man to keep up with. He was always a rugged individual with no time for nonsense. He impressed most who knew him as stern. Bob Rice knew better, for he spent hours with his grandfather in a workshop behind the house. There Henry Rice engaged in many projects and hobbies, including building metal, electrically powered toy boats, which he sold to toy stores. He also tinkered with Boothbay engines. He lived thus until 1959. Bob Rice doubts that he will see the likes of his grandfather again.

THE MARBLEHEAD MACHINE COMPANY

Although we do not know the name of the engineer who designed the first two-cycle Acadia engine for the entrepreneurial W.T. Ritcey, the man who designed Acadia's first four-cycle engine was named Arthur E. Colchester. He was born on his grandfather's estate near Windsor, Nova Scotia, in 1874, the son of a mining engineer who perished of yellow fever in South America at the age of 40. By then, Colchester's grandfather had lost most of a fortune amassed by selling a steel-making formula to Bessemer in Pittsburgh. The elder Colchester mortgaged his estate to subsidize a determined but vain search for copper in Newfoundland. Years later, with the help of a Geiger counter, copper was discovered one foot away from the deep shaft that had been Colchester's ruin. That at least, is the way the family remembers things.

Rather early in life, Arthur Colchester developed an intense interest in steam engines that was surpassed by an even deeper interest in gas engines. He decided to move to Boston and seek a career, but, instead of attending M.I.T. as his father had wished, he studied drafting and then took a job in a New Bedford boatyard. There he learned all the intricacies of gasoline engines, taking them apart, repairing them, machining parts as needed, and often making drawings of castings and other pieces.

By 1905 or 1906, Arthur Colchester had learned enough to go into business for himself. He moved to Marblehead and established a fine machine shop among the boatyards on Lee Street. "Gasoline Engines Repaired & Adjusted," said the sign. "Automobiles Repaired." Colchester, wearing a suit and bowler hat, was photographed in front of the building with his wife, small son Herbert, and his accountant.

In this shop, Colchester repaired engines for the many yachts and fishing vessels that frequented Marblehead's busy harbor. He also designed and built his own two-cycle one-lunger and installed it in an 18-foot boat, which he used for family outings and as a promotional device. Apparently, he was satisfied on both counts and, after a trip around the point to Salem caught him in a sudden blow, he recorded that his engine drove him home "without missing one explosion."

Nobody remembers, anymore, just how many engines Arthur Colchester built and sold. There may have been fewer than a half-dozen; there may have been more. He was only one of the countless machinists around the country who played a small part in the great change from sail to power, but who have been all but forgotten.

Gerald Smith knew Colchester. He bought a 3 hp engine from the machinist and installed it in one of George Chaisson's graceful dories. In this boat, Smith went out each day to set trawls around the rocky islands off Marblehead — until, one day, the wind blew so hard that he could barely make it home. "I had to sort of tack," he said, "like I was in a sailboat."

Smith replaced the little Colchester with a 5 hp Mianus. Later, he sold the Colchester, and it was immediately put into a Swampscott dory by its new owner and used for work. Smith remembers that Colchester's engines were entirely satisfactory, except that the washers that seated the igniter would eventually wear and allow compression to escape, a not-uncommon failing on make-and-break machines.

"I'd take a stick," said Smith, "and push that cussed igniter arm in there and

Arthur E. Colchester, 1903.
(Courtesy Hilda Weston)

get a few more rpm that way." He also tightened down the grease cups, fore
and aft, to minimize further compression loss out along the main bearings.

Even during the years when he owned his own company, it was apparent to
Colchester's family that he had little skill or, perhaps, interest, in the business
side of things. Customers were seldom prompt in their payments, and Colchester was inevitably more enthusiastic about his latest engine project than
about collecting bills. Needing a more reliable income to support his wife and
three children, he closed Marblehead Machine Company in 1912 and moved to
Boston. There he worked for the robust firm that was Murray & Tregurtha,
performing a variety of jobs, from design and engineering to repair and installation. He stayed with the company until shortly after the end of World
War I. At that time, business began tapering off and Murray hired a new
manager who happened to be a trained engineer.

"Our father didn't have higher math," remembers Colchester's daughter,
Hilda, "and he and the new manager did not get along."

In 1919, Colchester took his wife and family back to Nova Scotia. They
settled in Bridgewater in a house just behind W.T. Ritcey's and Colchester
began work on a new four-cycle engine for Acadia. This was, by all accounts,
quite a splendid piece of work, a valve-in-head four-cylinder, each cylinder
head an individual gray iron casting. The upper crankcase was cast of
aluminum, the lower portion of bronze. The camshaft was notable in its day,
as it would be even today, for having hardened and ground rollers, a feature
one might expect to find only in a modern racing engine. Lubrication was provided for by a gear-driven bronze oil pump. The machines were equipped with
dual ignition systems, a battery arrangement using Bosch components and an
Atwater Kent magneto.

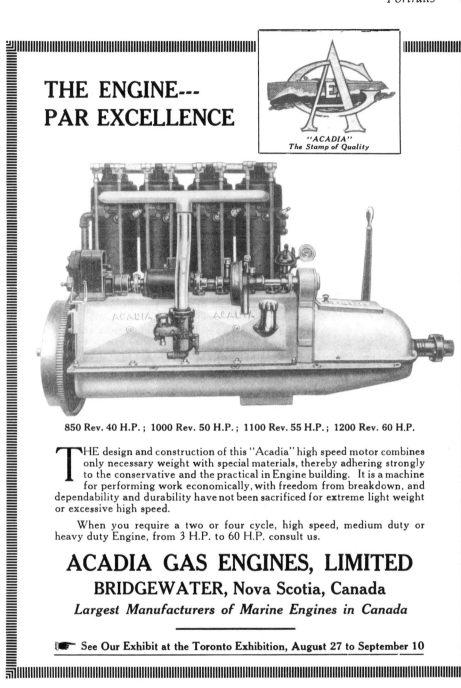

Ritcey installed two of these engines in his sharp-bowed, hard-chine speed-boat, the *Migrace*, and immediately possessed the fastest boat on the La Have River. He used to take his daughters and Colchester's children for rides.

"Those were very reliable engines," remembers Stan Forbes, who, soon after going to work for W.T. Ritcey, was given the job of designing a ring gear and electric starter for the machine. "If anything happened to one cylinder, you would just disconnect the spark plug and carry on with what was left." Eventually, the engines powered Ritcey's boat on a cruise around Nova Scotia. On the trip, *Migrace* occasionally covered 180 miles between morning and evening. She was fast, but, by most accounts, wet and uncomfortable. With her narrow bow, 32-foot length, six-foot beam, and V-bottom, she had never been intended as a sea boat.

The engines' performance on this 1,000-mile cruise was duly noted in the press, as were details of Acadia's production techniques. The company was then capable of producing some 3,000 engines per year. In the tradition of two-cycle engine manufacture, parts of this four-cycle were made right at the Acadia shops, creating a problem that stalwart two-cycle advocates had always foreseen. There were so many more parts to a four-cycle engine that manufacture was comparatively costly unless mass-production systems were employed. Acadia ground its own camshafts, cut its own timing gears, and, of course, fitted everything together by hand. The result was a stout engine that became a favorite of the Royal Canadian Mounted Police. But it was expensive, too expensive to become an important factor in the marketplace.

Colchester's daughters remember that, once the job of designing and perfecting this engine had been completed, his agreement with Ritcey ended. In 1921, the family moved back to Marblehead, where Colchester's children attended school. He commuted daily to a job in a Neponset boatyard. "Somehow," said his daughter, Hilda, "Father was never quite settled in things."

After his return to the United States, Colchester designed and patented a valve that required no springs or tappets to operate, apparently a precursor of the desmodromic cylinder head used 30 years later in certain Mercedes-Benz racing cars and a number of motorcycles. He tried to interest General Motors President Alfred Sloan in his ACOL (for Arthur Colchester) valve. Sloan may have found the device intriguing but was not about to finance the retooling of every engine in GM's line to accommodate a new valve. When he couldn't sell the valve, Colchester set up a new machine shop. It was destroyed by fire and there was no insurance. The Depression found him living in Dorchester, working in his cellar on a rotary engine that ran but developed little power, perhaps because the technology of the day was not up to producing adequate seals for the rotor's tips.

On projects such as these, the visionary machinist spent most of the money that came his way. When he died at age 85 in 1959, he was still dreaming about engines, but he had never made a financial success of any of his ventures. Nothing seems to survive of his work except, perhaps, for a lone builder's plate from one of his old one-lungers. The drawings he made for the four-cycle Acadia engine were lost or destroyed. There remains only the building in

Marblehead, where Colchester's machine shop once was housed. But it has long since been turned into a dwelling. Behind it, where once there were boatyards, the buildings have all been converted to homes. Expensive cars are garaged where the boats used to be stored. The only man in town who remembers Arthur Colchester is Gerald Smith.

J.W. LATHROP & COMPANY

James W. Lathrop was a needle peddler who earned a fortune building marine engines. He made his way from Worcester, Massachusetts, to Mystic, Connecticut, in pursuit of customers for his needles. Eventually, he formed a partnership, but one day the partner took all their money and left. Later, J.W. would admonish young men about the dangers of being in partnership with anybody.

In Mystic, Lathrop became friends with a man named Gallup who owned a launch. It was powered by a Palmer one-lunger, and the two men frequently enjoyed excursions on the Mystic River. J.W. was so impressed that he even wrote the Palmers an appreciative letter about their engine. On one of these river jaunts, the Palmer developed a problem with its igniter. This so frustrated Gallup that when they returned he upended a sugar barrel and, on Baptist Hill on the west side of Mystic, drew a sketch for an improved make-and-break sparker. J.W. was a fascinated bystander.

"So what did they do? They came down to the riverside to a little shop and made up a whole engine. They sold it and got enough to buy a lathe and built two more engines. From then on, they were in the engine business."

That is how James W. Lathrop described the beginning in 1897 of J.W. Lathrop & Company to Edward R. Welles, Sr. Welles worked at Lathrop's for 30 years. He joined the company in 1916, after his father's sawmill burned down.

There seems little doubt that the first Lathrop borrowed rather heavily from the Palmer, as others did, but J.W. soon developed his engines so they possessed their own unique character. Like Palmer, his company was among the longest lived of all those that began to build marine engines around 1900. By the time President William McKinley had held office for a year, conditions at J.W.'s little works were described as booming, and Lathrop, together with Palmer, was considered among the industry's great pioneers. The local newspaper enjoyed noting Lathrop sales and in February 1901 reported the shipment of two 5 hp Lathrops to Seattle, Washington, another pair to Jacksonville, and two more to Bath, Maine.

For many years, J.W. oversaw every detail of his factory's operation. It was a robust enterprise. Welles recalls that, when he joined the company, 90 men worked there. At first, Welles ran a drill press. He earned $9 for his first week's work, then received a raise to $12, and was assigned to operate a machine cutting keyways for shaft couplings. These he made by the barrelful. Next he worked in the finishing gang that completed engines — 10 to 14 at a time — in one corner of the factory. He soon became a foreman. When he suggested to

J.W. that money could be saved if a jig were built so manifolds could be mass produced instead of built individually for each engine, the suggestion was acted upon immediately.

Thanks to his engine-building company, James W. Lathrop became a wealthy man. He owned several beautiful homes in Mystic and once boasted to Welles that he had money in 10 different banks and owned two of them. His son, Walter, designed the company's first four-cycle engine and gradually assumed enough responsibility so that J.W. could go to Florida in the wintertime.

J.W. Lathrop died in 1935. His son carried on the business and was followed, in turn, by J.W.'s Yale-educated grandson. Gone, however, was the firm hand and intimate knowledge with which J.W. ruled his domain. Old-timers looked back and knew the flavor of the company had changed forever, not necessarily for the better. After a succession of owners in the mid-1950s, the company ceased to exist a few years later.

Welles finally left Lathrop in 1954. He set up his own shop and put to good use his years of experience. Today, his son presides over the shop, where, on occasion, a part for an old Lathrop still turns up, as if by magic, when a collector needs it. A big Lathrop one-lunger stands just inside the shop door. Like all the Lathrop two-cycles, it is painted red. It is ready to run at the closing of the switch.

THE FAY & BOWEN ENGINE COMPANY

It was during the 1880s that industry began to appear in the lakeside town of Geneva, New York. There were stove factories, canneries, foundries and optical companies, companies to make knives, cereal, and malt. A branch of the Erie Canal served the town, and so did four railroads. These transportation links were among the reasons why, in 1904, the Fay & Bowen Engine Company moved to Geneva.

"We consider," said the proprietors, "our location unsurpassed by that of any manufacturer in our line."

The company could launch its boats into the still waters of the feeder canal and then deliver them to the Great Lakes, the Thousand Islands, or the East Coast. Customers were urged to take one of the many trains serving the town, stay in one of its three hotels, visit the Smith Opera House, where Broadway plays, Caruso, and other luminaries often were featured. After that, and perhaps a stroll through Geneva's quiet streets lined with Victorian and Greek Revival houses, the buyer could take the helm of his new Fay & Bowen boat and motor off to years of happy cruising.

That's how it was in Geneva in those days, when the Fay & Bowen shops stood at the head of Seneca Lake and gleaming boats slid into the canal and John Philip Sousa played at the opera house. The good times lasted for nearly three decades.

Walter L. Fay and Ernest S. Bowen formed their first company in Auburn,

The Fay & Bowen works, Geneva, New York, circa 1905. (Courtesy Geneva Historical Society)

New York, in 1895. They manufactured bicycle spokes and spoke nipples. The men's partnership was a happy one, made at a happy time, since substantial fortunes were then being won in the bicycle industry. When the two sold their factory on July 4, 1900, they each garnered a tidy reward. This money was immediately reinvested in a new company, one devoted to building engines. Bowen, a Cornell graduate engineer, had designed and built a two-cycle marine engine.

Looking back on the company's formative years in a 1917 interview with *Power Boating*, Walter Fay said, "We never built an engine that didn't run successfully. Nothing we put out was ever scrapped and we still hear from owners of machines we built in those early days." The company built stationary engines, too, and they were just as durable as those installed in boats. One of them was used for years by a local butcher to grind meat for sausage.

The very first Fay & Bowen engine was extensively tested in a 25-foot boat on Owasco Lake during the autumn of 1900. Once it had proved itself, manufacture was begun in a small brick factory at Auburn's Big Dam. Engines were assembled on one floor, painted and crated on an upper story, and lowered by fall and tackle to waiting trucks, which transported them two miles to the company boatshop. After the engines were installed in newly completed hulls, the vessels were tested on the lake. At first, most of the boats were fantail launches, but Fay & Bowen soon trended toward torpedo-stern models with flat underbodies in their aft sections.

"We had scored a tremendous success with this type," said a brochure, "and all of our stock boats are built on these lines. In standard models the motor is generally placed amidships and our installation is such that there is ample room to pass from bow to stern, the exhaust leaving the engine at 45 degrees and passing entirely out of the way under the floor of the boat."

In addition to launches, speed and "cabin" boats were built. They were constructed of white oak frames and cedar or cypress planking, copper fastened. White oak was used for the sheerstrake. A high standard of finish was maintained on the company's boats. Fay's daughter, Mildred, remembered that they "were very expensive, even in those days." They were especially popular in the Finger Lakes and Thousand Islands, but agencies eventually were established in some 22 countries.

F & REED
LATHE

Machine shop, Fay & Bowen Engine Company, Geneva, New York, circa 1910.
(Courtesy Geneva Historical Society)

It was apparently toward the end of 1902 that Messrs. Fay and Bowen began planning a move from Auburn to Geneva. During the winter of 1904, they set up the business in an impressive array of buildings supplied with electrical power by a three-cylinder, two-cycle Fay & Bowen engine. The company prospered and soon became a large local employer. When times were good, just over 100 people worked there.

One of these employees was Fred Breitfeld, who became an apprentice in the Fay & Bowen boatshop when he was 15 years old in 1912. He began by sweeping floors, clearing work areas, making plugs for fastener holes, and mixing putty for the boats' seams. Eventually, he worked in most of the departments. "I worked all over the place," he said. "I worked in the pattern shop for quite a few years. I helped make patterns for the marine motors."

Like others in those times, he worked 10 hours each day, beginning at 7 a.m. and ending at 6 p.m. He remembers that it was "quite a while before we had Saturday afternoons off."

By the time America entered World War I in 1917, Fay & Bowen had grown to comprise some 10 or 12 buildings, including a 42-by-45-foot installation shop equipped with cranes to lower engines onto their bearers, an upholstery shop, a boatshop, machine shops, stockrooms, and sheds where townsfolk could store their boats. In 1916, manufacture of a four-cycle engine commenced when the trend toward ever-larger vessels became apparent. "More of the other companies were going into four-cycle motors," remembered Fred Breitfeld, "Chris-Craft, Dodge, and lots more."

Ernest Bowen was working on a design for a four-cycle engine — to be used in cars as well as boats — when he died of typhoid fever at the age of 53, in 1913. It was his basic design, however, that was used for the production

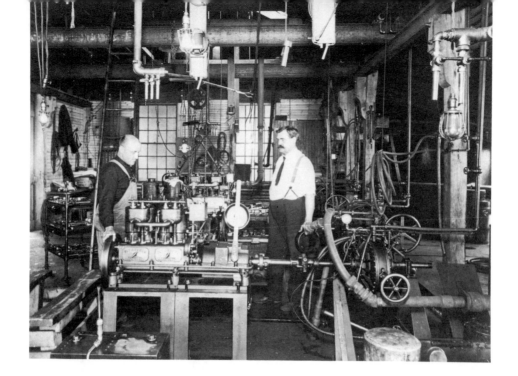

In the test house a Fay & Bowen four-cylinder, four-cycle T-head engine is ready to go. Circa 1915. (Courtesy Geneva Historical Society)

engine. Like the two-cycle, the "four" was characterized by the finest materials, oversize bearings, and superb workmanship. These were T-head engines in which the inlet and exhaust valves were arranged on opposite sides of the head, allowing the valves to be of particularly generous dimensions. No engineer of that epoch could have remained uninfluenced by the successful use of T-headers in Wilhelm Maybach's first Mercedes, in 1901, or — closer to home — the wonderful T-head engine designed in Trenton, New Jersey, by Finley Robertson Porter for the Mercer automobile.

During the war, Fay & Bowen received a contract to build the Curtis HS-1, an elaborate seaplane powered by the Liberty V-12, then being mass produced by several automobile makers. The company built the seaplane hulls and shipped them to Buffalo. Fred Breitfeld was in charge of production. "They took them to Buffalo and put the wings on and the motors in," he said. "What they were after was Fay & Bowen's experience building boats."

Walter Fay retired in 1919 and, although he occasionally visited the works after that, and although the company was left in capable hands, things began to change. During the 1920s, when buyers placed ever-increasing importance on speed, Fay & Bowen failed to keep pace. It is not certain whether this was a policy decision — a refusal to sacrifice the smooth ride and sturdiness of substantial, round-bilged boats for hard-chine runabouts — or a lack of adequate capital to finance the necessary tooling. The onset of the Great Depression was the end of Fay & Bowen. The fine buildings were taken down and sold. Eventually, the old canal was filled in and a road was built. Now, a highway runs right through the site where the old company once stood.

Every great principle founded on truth, whether of morals or mechanics, has been obliged to fight its way to the front, slowly but surely, superseding the knockers and blunderers and those sincere but mistaken. Such has been the history of the two-cycle type of marine engines.

C.T. Wright Engine Co., Greenville, Michigan, 1911

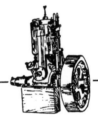

3

The Art and Science
of the Two-Stroke

WHEN THE FIRST MANUFACTURERS of two-cycle engines placed their wares on the market, they touted the type's simplicity in comparison with its four-cycle competitors. This emphasis on simplicity went on for years, even as the good fight was being lost and four-cycle engines became predominant in automobiles and boats. After a while, two-cycle-engine builders' literature began to sound a bit desperate, a plaintive cry to which few cared to listen. Companies had to battle against increasingly functional four-cycle engines and against an image — held by a distressing number of people — that, simple or not, a two-cycle engine just was not as good as a four-cycle.

"The reputation of the two-cycle motor has been injured," said a Syracuse Gas Engine Works brochure, "by engines designed and built by persons having no adequate conceptions of the requirements of the case. These poorly made motors are no more a real argument against two-cycle construction than to say that a chronometer cannot keep good time because a dollar watch does not."

Because of problems with some early two-cycle engines, opinions began to be formed in many quarters that could realistically be termed prejudice. Considering these feelings in 1906, an engineering writer concluded: "The two-cycle engine is said to be the pet aversion of those who have built four-cycles only, or who have simply dabbled in two-cycles. Why? The answer is that almost anyone with a little skill in proportioning and a knowledge of steam engine design can design a four-cycle, but it requires experience, often dearly bought by expensive experiments, to design a two-cycle, especially a high-speed one."

There was, in fact, more to the two-cycle than met the eye. Simple in appearance, the successful two-cycle was not simple to create from a blank sheet of paper. Although it was true that the engines had fewer parts than a four-cycle, such parts as there were had to be properly designed and constructed. The real secret to a successful engine was hidden from the eye. It lay within the cylinder casting itself, and only a cylinder with properly proportioned and located ports could be satisfactory. The engine's power, economy, and efficiency largely depended on these ports.

"The most important things from a designer's standpoint are, as a rule, ports, port location, and port areas," engineer E.W. Roberts told a gathering of the Association of Licensed Automobile Manufacturers in 1908. Roberts was an Ohioan who had by then designed about 92 engines and was an acknowledged engine expert.

Acadia marine engine, 4 hp, two-cycle make-and-break.

An early (circa 1897-98) two-cycle one-lunger built by Murray & Tregurtha in South Boston. At 600 pounds, this 4 hp model was at least twice as heavy as engines of similar output that were built only four or five years later. Still, the company could rightly proclaim the machine was "very light and compact" for its time. The engine is unusual for its automobile-like crank starting system and flywheel enclosed within a polished brass shield. Ignition was make-and-break. "There are no lamps or burners, and no matches are required," said the manufacturer, referring to the then-prevalent hot-tube systems.

Palmer — igniter in "make" position.   *Palmer — igniter in "break" position.*

Lathrop 7 hp — igniter in "make" position.   *Lathrop 7 hp — igniter in "break" position.*

PORTS

Two-cycle engines were of two basic types: the two-port and the three-port. In both cases, the operating principles were the same. The accompanying illustrations show the operation of each type.

The air/gas mixture was admitted to the crankcase on the piston's upstroke when a vacuum was created. On the downstroke, the mixture was compressed within the crankcase, or base, and forced through a transfer port — also called a bypass port — to the inlet port and combustion chamber. The burned mixture was eliminated through the exhaust port on the piston's downstroke.

The basic difference between the two-port and the three-port engines was the means by which the air/gas mixture was admitted to the base. On the two-port engine, it entered through a one-way check valve mounted low on the base. The purpose of the check valve was to prevent the mixture from being forced right back out of the base instead of into the bypass port. On the three-port engine, the check valve was eliminated. The mixture was drawn in through a third port in the cylinder wall that was uncovered by the piston near the top of its stroke.

Since two- and three-port engines were on the market at the same time, manufacturers immediately began promoting the virtues of whichever system

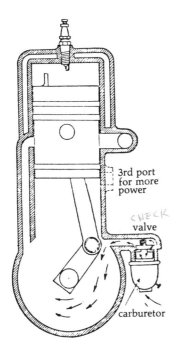

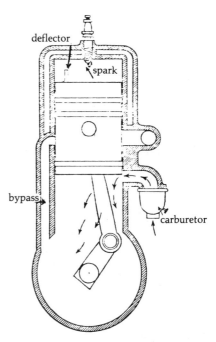

TWO-PORT MOTOR
Air entering base from carburetor through check valve. Air enters during entire upstroke of piston.

THREE-PORT MOTOR
Piston nearing top of stroke; port uncovered by lower end of piston; mixture entering base from carburetor; explosion taking place above piston; both ports leading to cylinder closed.

These drawings, from a 1911 issue of Yachting, *show the difference between a three-port engine — with its piston-controlled inlet — and a two-port, in which the inlet is governed by a check valve.*

they used and pointing out alleged deficiencies in the other. Northwestern Motors of Eau Claire, Wisconsin, built two-port engines and claimed they were the easiest to start. "Especially is this so after the crankshaft has worn its bearings [and the pressure that could be generated within the base was thus lowered]. The three-port engine has the disadvantage of forming a vacuum before the port opens. The vacuum is formed below the piston upon its upward stroke, and detracts from the power produced, and is sure to show in the bearings sooner or later, instead of holding the vacuum until the port is opened, air will be drawn through the bearings, reducing the vacuum so there will not be enough suction when the port does open, which causes loss of power . . . . The three-port engine has been tried and found wanting . . . . The two-port engine is the simplest of the two types, and simplicity is what both manufacturer and buyer should aim to secure."

In contrast with this viewpoint, Syracuse Gas Engines pointed out that the check valve of the two-port engine was dependent upon a spring to close it, and Syracuse noted that springs could break. Further, the company noted cor-

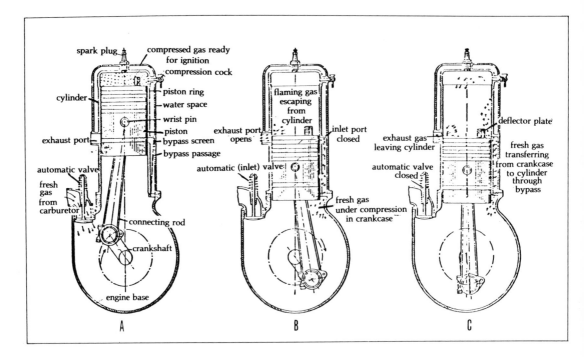

The labeled diagram shows three positions (A, B, C) of a two-port, two-cycle engine with the following labels:

A: spark plug, compressed gas ready for ignition, compression cock, cylinder, piston ring, water space, wrist pin, piston, exhaust port, bypass screen, bypass passage, automatic valve, fresh gas from carburetor, connecting rod, crankshaft, engine base

B: flaming gas escaping from cylinder, exhaust port opens, inlet port closed, automatic (inlet) valve, fresh gas under compression in crankcase

C: deflector plate, exhaust gas leaving cylinder, fresh gas transferring from crankcase to cylinder through bypass, automatic valve closed

A good illustration of the two-port, two-cycle in action. (From Motor Boats and Boat Motors*)*

rectly that if the check valve was "wrong or improperly adjusted, the incoming gas does not reach the cylinder at the proper time . . . or, if not perfectly tight, a leak occurs seriously impairing or entirely destroying the efficiency of the motor."

Rhetoric aside, good engines of both types were built. However, in too many instances, it was the two-port with its check valve that performed poorly, and the two-port predominated during the early years of two-cycle development. An editorial in *The Horseless Age* in May 1907 noted:

> Very many people who have not followed the situation closely still evidently regard a two-cycle engine as necessarily the original type that has been so long used in power boats, namely, an engine with piston-controlled inlet and exhaust ports, a closed crankcase into which gas is drawn from the carburetor through an automatic suction valve, and there compressed and passed to the cylinder through a by-pass pipe or cored passage.
>
> The general public has even yet hardly become familiar with the three-port type of two-cycle engine, which has shown itself so practical in the hands of resolute manufacturers. The three-port arrangement, in which communication between the carburetor and the crankcase is made by a piston-opened port instead of by a suction-opened valve, was the first successful variant made from the widely used marine and stationary type of this form of motor.

The inherent fallibility of the two-port engine lay in the action of the check valve that tended to limit rpm, for, as speed increased, the valve's action could become somewhat uncertain. It did not always seat itself squarely and thus permitted some of the mixture to be blown back around it instead of remaining in the base where it belonged. Some theorists speculated, too, that because the mixture was drawn in during the entire upstroke of the piston, less efficiency was achieved in the two-port than in the three-port. On the latter, vacuum built up as the piston rose, and, when the third port was uncovered, the mixture rushed in with a puff. This was generally believed to permit more positive action as speed increased.

After all of his experimenting and design work, engineer E.W. Roberts came out strongly in support of the three-port engine. He preferred it because of its greater simplicity — no check valve — and because he learned that it gave better results than a two-port at speeds around 600 rpm or better. The three-port, two-cycle engine reduced the internal combustion engine to one of its simplest forms and came to be considered the classic example of two-stroke design.

Although two-cycle engines are conveniently classed as either two- or three-port types, there were variations to both. Occasionally, a manufacturer would designate his engine as a two-three port, an example of which is illustrated. By directing fresh mixture into the bypass port, heated by the previous explosion, better vaporization of fuel was achieved. Again, however, a check valve was required to keep the freshly admitted mixture where it belonged. Advocates of such engines also claimed that the additional suction, created because the mixture was drawn in over a longer period than was possible with a three-port, increased power. This idea was probably overstated and was an obvious and unwarranted rejection of the "puff" theory. One test of the suction generated by a three-port two-cycle found suction at the inlet so strong that it could hold an eight-pound box against the opening.

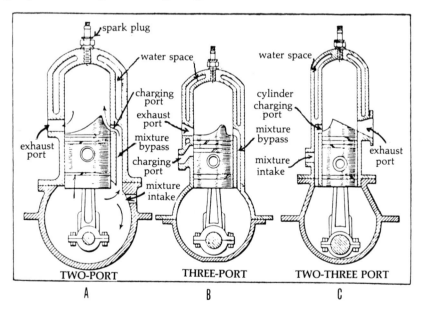

Types of two-cycle marine motors.

Sometimes, an additional air-inlet port was added to the cylinder and counted as a third or fourth port, depending on who was doing the counting. The port permitted a charge of fresh air to enter the combustion chamber just before the mixture's entry from the intake port. The auxiliary air port was intended to increase power somewhat by cleaning the mixture and to help drive out the hot, burnt gases of the previous charge. This reduced the chance that those gases would escape back into the base and cause a backfire. The air port was operated either manually, by sliding open a covering plate, or automatically.

PORT LOCATION AND AREA

The most basic considerations for the designer laying out a new two-cycle engine were — after bore-stroke dimensions — the location and area of the ports, and port timing. The latter was expressed in degrees of crankshaft rotation. As a general rule, the inlet port to the base was left open as long as possible without running the risk of blowback through the carburetor. On a single-cylinder three-port, the port generally began to be uncovered 44 degrees before the piston reached top dead center and remained uncovered for the same period as the piston started downward. As the piston descended, the exhaust port was uncovered for 58 or 59 degrees on either side of bottom dead center. The inlet port to the combustion chamber opened a few degrees after the exhaust port and closed a few degrees before the exhaust port was sealed. A port-timing diagram is included in this chapter.

Port timing was crucial to the successful two-cycle engine. The exhaust port, for instance, could not be opened too soon, or power would be lost due to lessened pressure within the combustion chamber. However, if the exhaust port opened too late or was not large enough, hot gases might rush back into the cylinder inlet port, through the bypass and into the base. There it could

A valve timing diagram showing the position of the crankshaft at the opening and closing of each of the three ports and the time during which the ports remain open. (The Horseless Age)

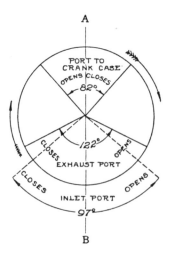

mix with the fresh fuel and cause an explosion. Explosions in the base were the bane of some two-cycle engine designers. In extreme cases, these explosions could blow the grease out of the main bearings, causing the base to lose compression and stopping the engine.

Laying out a two-cycle's ports was a tricky business. Ports of a certain size that worked well at one speed range often proved unsuitable when speed was increased. During the early days of two-cycle development, all designers, whether professionally trained or not, worked mostly by trial and error. Engineers experienced with four-cycle engines faced the reality that, in a two-cycle, they had less than half the time to achieve the same inflow and outflow of gases. The inlet valve of a four-cycle was open from 200 to 210 degrees, compared with the inlet port of a two-cycle's approximately 90 degrees. E.W. Roberts, in the paper he read to the Association of Licensed Automobile Manufacturers in 1908, advised: "Always lay out the angles on the crank circle and work to the points these angles will give, and remember that it is a case of time, not distance."

Because the two-cycle engine cut in half the time required by a four-cycle to perform the same combustion process, it was generally believed that the area of the intake port, for example, had to be twice as great as a comparable four-cycle's intake valve. In order for fresh mixture to enter the two-cycle quickly, and in order for burned gases to be expelled quickly, inlet and exhaust ports were as large as possible. How *large*, of course, was the dilemma. In 1907, the mechanical laboratory at the University of Michigan conducted tests to determine the proper dimensions of ports in a 4½-inch-bore, 5-inch-stroke, three-port engine.

These tests were exhaustive and probably exhausting, since the engine was cast with ports as small as practical, and these ports were then enlarged gradually over a series of 12 test periods, $\frac{1}{16}$ inch at a time. This process was continued until, on the thirteenth test, the chisel being used to enlarge the ports slipped and pierced the water jacket at a thin spot in the casting. The tests showed that horsepower increased with an increase in port area — slowly at first, then more rapidly, and finally more slowly again as a point of diminishing return was reached.

Among other things, the Michigan tests showed that port size had to be based on a decision that balanced horsepower against fuel consumption and excessive loss of heat energy. They found, for instance, that power increased constantly as the exhaust port was increased in size, but that efficiency dropped because burned gas was being expelled at greater temperature. The same could be said of the transfer port. The inlet port was found to increase power for a time, as it was enlarged, but then to decrease power.

The report concluded "that choosing the correct size of ports is entirely a matter of weighing the conditions under which the engines are to be used and deciding where the balance between efficiency and horsepower shall be struck." As a result of its scientific approach, the report suggested these port dimensions for a 4½ x 5 engine: inlet, 2⅛ x ⅝; transfer, 2¼ x ⅝; exhaust, 2⅝ x 1³⁄₁₆.

THE ROLE OF THE BASE

Because the initial compression of the air/gas mixture in the two-cycle occurred in the crankcase (or base, as it was then commonly referred to), the base had to be sealed tightly. If there were leaks, the suction generated by the rising piston was lessened, and pressure generated by the descending piston would also be diminished. The base was a casting, the interior of which was just large enough to clear the crankpin as it rotated. Often there was little more than ⅛-to ¼-inch clearance provided for the bottom end of the connecting rod. It was realized rather early that discs attached to the crankshaft webs could increase pressure in the base. Most companies, however, found attaching such discs more trouble than they were worth in an era when crankshafts were seldom even balanced. They relied, instead, on making the base itself a close fit to the rod.

The primary place for pressure generated within the base to be lost was out along the crankshaft, through the main bearings. To reduce leakage, the bearings were made rather long. They were lubricated by grease cups, and the grease, forced down into the bearings, acted as both lubricant and sealant. Pressures within the base ranged from 4½ to 6 pounds, although engines intended for racing might be pushed to 8 pounds. In practice, most two-cycle engines had enough wear along the bearings so that they were seldom really tight. In fact, when installed in open boats, the engines commonly gathered a certain amount of water in the base after a heavy rain or when the bilgewater rose high enough to cover the bearings. The base had a petcock intended as a drain for a gas-flooded engine, but most owners used it more often to drain off water before starting.

In the winter, water that entered the base was liable to freeze. Here is what Bob Merriam said about starting such an engine. "I could get the flywheel to move just a little bit. The idea was to trip the igniter and get the engine to fire one time. It would go 'boom' and blow all that ice right out the exhaust pipe, and then away you went."

As the main bearings became worn, they became ever more prone to leakage. If they were one-piece, circular bearings, they then had to be replaced. If the bearings were made in halves, a shim could be added between the base and the top half of the engine and tightness restored. Often these shims were merely a few thicknesses of waxed or greased paper. Sometimes they were made of metal.

Occasionally, a company took a different approach to the problem of sealing the main bearings. Northwestern devised a method of tightening the bearings without dismantling the engine or replacing the bearing itself. It was described by the company like this: "A groove is cored into the main journal, into which is fitted a half steel collar split on either side . . . . A ring of flexible cylinder packing comes into direct contact with the crankshaft. Wear, when it appears, is taken up by a quarter turn of a small set screw which is placed at both top and bottom of journals. These screws only require attention once or twice during the season." Looking at Northwestern's illustration of the device, however, one wonders if the set screws applied pressure evenly around the crank or if there was a danger that they could be tightened too much. Still, the motors were in production long enough for the adjustment to prove its worth and were one more example of that ingenuity and individuality displayed by those who designed and built two-cycle marine engines.

VAPORIZERS AND CARBURETORS

Until 1910, the vaporizer was the chief means of mixing air and gasoline and delivering the combustible mixture into the engine. These simple devices, in use before the turn of the century, were also referred to as generator valves or mixing valves. Gasoline entered the vaporizer along a needle valve. It was stopped from flowing out into the vaporizer itself by the edge of a check valve. When the valve was opened by the suction of the piston, air flowed past the needle-valve orifice, picked up the fuel, and entered the base as a mist. Sometimes the mixing valve was explained to laymen as akin to a perfume atomizer. When the piston reached the top of its stroke, suction ceased and the valve was returned to its seat by the pressure of a coil spring.

The simplest vaporizers' only adjustment was the needle valve itself. More sophisticated models permitted valve lift to be regulated, and some had a throttle plate as well. Despite these embellishments, the vaporizer remained basically a one-speed instrument. It was seriously compromised by tiny particles of dirt that quickly clogged the needle valve or kept the check valve from seating tightly. The vaporizer was also affected by the fact that, as the gas tank emptied, the pressure at which the gasoline flowed varied, and by the rolling of the boat. Since there was no automatic form of compensation for such inevitabilities, the vaporizer required adjustment by a sensitive hand.

That the vaporizer was still in use in 1910 on some engines is a good indication of the problems engineers confronted when trying to improve upon it. However troublesome the vaporizer might have been, it was for a long time less odious than a carburetor. "For a marine gasoline engine," said a *Rudder* article in 1900, "the vaporizer is usually to be preferred to a carburetor . . . . Carburetors are wasteful of fuel in that they vaporize its lighter constituents, leaving a useless residue which has to be thrown away. They are also very sensitive to changes of temperature, and in extremely cold weather, it is necessary to heat them or the air that passes through them, in order that they may work properly."

During the first years of the new century, engineers struggled to overcome these complaints. They realized that once the carburetor was perfected, its float chamber — which automatically kept level the amount of fuel in the instrument — would yield a device infinitely superior to the vaporizer. But carburetor development was slow. "The carburetor problem on a two-cycle motor is a difficult one," bemoaned one engineer in 1908. "We have had carburetor men at our factory time and again who confess themselves 'beaten.'

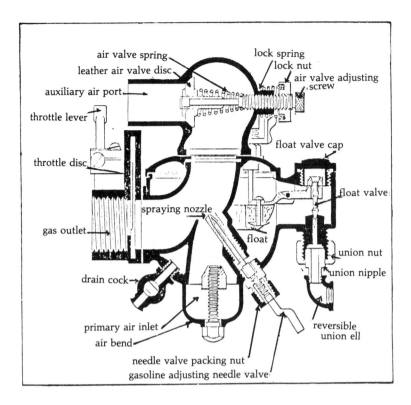

Schebler carburetor construction. This is one of the simplest forms that have been used extensively.

One thing I have found is that the initial air intake on a two-cycle carburetor should be much less than on a four-cycle . . . . Why this is, I cannot tell."

Gradually, however, the progress was made. It was realized that only perfection of the float itself would yield a basically sound carburetor. For years, floats had been made of cork. When the cork became fuel soaked, its efficiency was impaired and the engine's operation was then affected by the amount of fuel in the tank rather than the amount in the float chamber — just like a vaporizer. The cork float eventually was replaced by one of metal, usually brass. This was a great improvement, although a tiny leak in a metal float certainly caused much woe, for leaks of that sort were hard to detect.

Like the vaporizer, the carburetor was subject to clogging with dirt. It was recommended that fuel — which often varied greatly in quality — be strained either through a metal screen or, lacking that, a piece of muslin. Usually such advice was ignored, of course, especially by fishermen. Sometimes, an old felt hat might be used as a strainer, but, often as not, the gas was just dumped in. The tank itself frequently was made of a discarded fuel drum. Its only concession to filtering was to have the delivery pipe extend a bit above the tank's bottom so that sediments had someplace to go. The bottoms of those tanks frequently were covered with muck. Little wonder that both carburetors and vaporizers sometimes were clogged with the stuff.

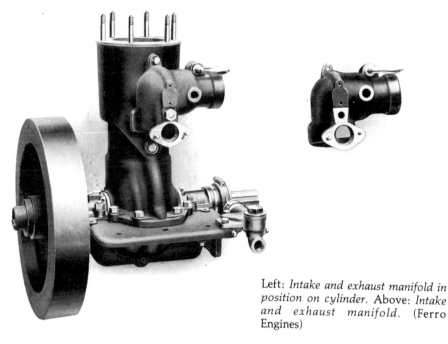

Left: *Intake and exhaust manifold in position on cylinder.* Above: *Intake and exhaust manifold.* (Ferro Engines)

Critics of early carburetors often noted that as temperatures dropped, so did the ability of the carburetor to deliver a gas/air mixture in gaseous rather than liquid form. In fact, the same problem existed with vaporizers. In cold air, gasoline does not vaporize as readily as it does in warm air. This problem was recognized long before the first carburetors were developed. It was suggested that an engine be warmed up before starting by filling the water jacket with boiling water, or by placing a hot brick or stone against the vaporizer or carburetor. As early as 1901, the Lozier Motor Company in Plattsburgh, New York, attempted to supply the vaporizer itself with air that was "warmed and dried in a special chamber or hot-air drum by means of exhaust gas from the engine."

Eventually, such tactics became standard practice. The Ferro engine, built by the Ferro Machine and Foundry Company in Cleveland, combined the intake and exhaust manifold into a single casting and mounted the manifold partway up the cylinder to take further advantage of heat given off by the water jacket. Said Ferro: "The heat of the exhaust gas and the constant circulation of the water keep the passage through which the ingoing gas is carried at an even temperature of about 78 degrees. This prevents the ingoing gas from condensing."

Although some form of heat became the accepted remedy to the problem of operating gas engines in all weather, E.W. Roberts experimented with at least one other method. This he described in *Yachting* in 1911. He installed a "honeycomb structure in the transfer port or bypass, through which the gasoline vapor and air must pass when going from the base to the cylinder.

This not only breaks up the cohesion of the atoms held in suspension, causing the mixture to pass over a surface of large area, but slightly raises the temperature and overcomes the chilling effect of the carburetor."

Perhaps because of the labor that was expended on developing practical carburetors, manufacturers tended to describe them in particularly glowing terms. Here is what the C.T. Wright Engine Company said in 1910: "It would be folly to spend a lifetime of thought and labor perfecting a gas engine and then fit it with troublesome accessories. That is about all there is to say regarding our carburetor. If it were not the best — if it gave the least trouble — if it did not do the work always without a hitch — you would not find it on a Weco engine . . . . Nothing more need be said about the carburetor. The studious life of an inventive genius is woven into the wonders of the Weco engine and the best accessories are none too good."

Few manufacturers were as smug as C.T. Wright. When Maine's Camden Anchor-Rockland Machine Company fitted a carburetor to its Knox engine in 1905, it described the instrument in some detail. It noted that the float did not leak, that the needle valve required no attention, and that the temperature of the air entering the carburetor could be regulated. A diagram showing the various parts was included. "The Carburetor," said the company, "is so simple that a child can operate it."

In theory, the speed of a carbureted engine was more readily controlled over a wider speed range than a vaporizer-equipped engine. At high speeds, the valve of the latter was likely to seat imperfectly, while at lower speeds, suction was not enough to reliably lift the check valve. Still, some manufacturers of two-port, vaporizer-equipped motors attempted to solve the problem. Lozier noted that "much of the difficulty encountered in the operation of gas engines has been in reducing speed. In most engines reduction of speed is attempted by changing the amount of gasoline or air, by means of separate valves. By this method the explosive quality of the gas is altered, and the engine frequently and unexpectedly stops. An attempt to regulate the speed in this manner is always attended with unsatisfactory results." Lozier mounted a butterfly-type throttle valve midway up the bypass port. Although such a throttle was difficult to make, it had the notable advantage of permitting gas to enter the crankcase at pressure undiminished by a throttle valve located in the vaporizer itself. It also helped reduce explosions in the base, serving as an obstruction to their progress. Perhaps that was its greatest value, because, throttle or no throttle, the vaporizer-equipped, two-cycle engine remained essentially a one-speed machine. Often such speed control as existed was achieved by adjusting the spark timing.

On a two-port engine, replacing the vaporizer with a carburetor did not greatly improve speed control. Such engines still needed a check valve added to the carburetor, and the valve remained a limiting factor on speed range. The advantage of a uniform mixture, however, remained. Three-port engines did have a reasonable speed range, and yachtsmen, perhaps, took some advantage of it. Fishermen, however, generally ran their engines flat out. A propeller was selected to give 500 to 600 rpm, and that, essentially, was that. Those who

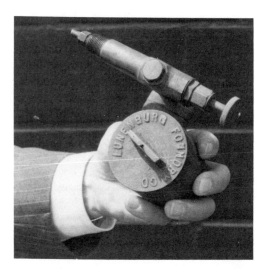

Fuel-oil carburetor.

seriously tried for lower speeds found 150 rpm to be about the slowest a three-port, two-cycle engine would run.

Carburetors, like vaporizers, were often mounted to a pipe that threaded directly into the base or cylinder. There was no intake manifold per se with this arrangement. On multi-cylinder engines, manifolds were commonly avoided by simply screwing one carburetor to each cylinder or base. It was understood that a single carburetor feeding more than one cylinder required a manifold that distributed the fuel mixture evenly to each. Experiments were conducted to determine how the manifold should be curved for best fuel delivery.

For those who wished to save money by running their engine on kerosene rather than gasoline, many manufacturers offered a kerosene carburetor as an option. This was fitted in addition to the regular carburetor and was fed from a separate tank. Usually these were simple needle-valve-controlled devices that metered kerosene drop by drop under slight pressure. More sophisticated kerosene carburetors also existed, and some carburetors were designed to deliver both liquids. The engine was first warmed up on gasoline, and then the switch to kerosene was made. Kerosene cost as little as one-fourth the price of gasoline and lacked the latter's potential for explosion — two reasons that fuel oil carburetors were popular with many fishermen.

IGNITION

Of all the one-lunger's components, none have remained so identified with it as the make-and-break igniter. When an old-timer talks about the engine he used to have, he usually calls it "an old make-and-break," or "an old make-and-break one-lunger," or "an old make-and-break two-cycle." But he gets the term in there somehow. Those igniters made a lasting impression on people.

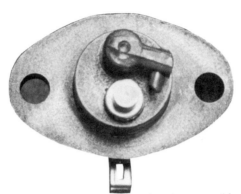

Igniter points showing movable and stationary electrodes.

Make-and-break igniters came into use during the early 1880s and, in many cases, replaced the hot-tube ignition such as the one Daimler and Maybach used in their first engines. Gone forever was the necessity of filling up a separate tank with fuel to supply the hot tube and then lighting the tube itself, rather like a Bunsen burner. A mistake, a flooded carburetor, gas flowing unchecked from a vaporizer — all could lead to spectacular fires. It's always nice to have a fire extinguisher handy when one is preparing such an engine for use.

Illustrations of igniter types are included here. The variety was limited only by the designer's imagination, which is to say that it wasn't limited at all. Still, all the igniters worked on the same basic electrical principle that a current, when interrupted, will jump across the ensuing gap in the form of a spark. Sometimes, igniters were called sparkers. In a make-and-break igniter, the spark occurred when a spring-loaded movable electrode was snapped suddenly out of contact with a stationary electrode.

The igniter was controlled by a rod driven off the crankshaft: as the rod moved up, it tripped a "firing pin," allowing the movable electrode to break contact with the stationary electrode. Sometimes the electrodes were mounted together on a single plate affixed to the side of the cylinder. In other engines, the movable electrode was mounted to the cylinder's side, while the stationary portion was screwed into the head where one might otherwise expect to find a spark plug. The low-tension coil that fired the igniter was connected to a group of dry-cell batteries yielding a minimum of 4½ volts. These batteries generally were mounted together in a box for convenience and usually lasted for a season. Sometimes a wet-cell battery or low-tension magneto was used.

The beauty of the make-and-break ignition system was that, being low-tension, it was impervious to moisture. Salesmen liked to dump a bucket of water over a demonstration engine to prove this, and the message was not lost on most witnesses. Said Stan Forbes of his years at Acadia Gas Engine: "We found we used 95 percent make-and-break. This was most practical for fishermen. You couldn't pour a bucket of water onto a jump-spark engine. They knew that when they went down to the boat in the morning, there would be salt spray on the wires and moisture from fog. When those engines would start up, if they were jump-spark, the fire would be flying over the outside of

Igniter, Atlantic engine. Note trigger and spring.

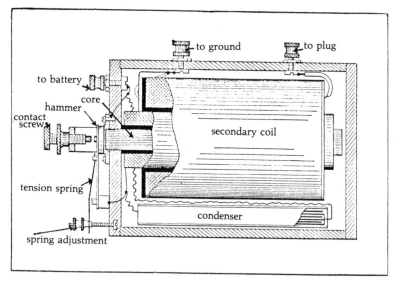

Part sectional view of simple induction coil, an important component of all battery ignition groups and sometimes used with magnetos.

the wire and she'd be missing. Jump-spark wasn't something you wanted if you were going out in all kinds of weather."

Along with its basic virtues, the make-and-break igniter had some basic problems. Mechanically, the system was complex when compared with a jump-spark system. The system of rods, arms, levers, and springs was subject to wear, which gradually changed the timing. Springs, in particular, tended to break. The igniter was further complicated by the addition of devices for advance and retard. Nor was it particularly practical to use a make-and-break system on anything but one- or two-cylinder engines. Nothing is impossible to find on old marine engines, however, and Weston Farmer — who took great joy in the old machines — discovered that Buffalo had built a four-cycle, four-cylinder make-and-break in 1908. The igniters were operated by an overhead camshaft. "All ye buffs of old engines," wrote Farmer in wonder, "gather ye and marvel!"

The jump-spark engine, with its high-tension coil and timer, possessed the virtue of mechanical simplicity at the expense of electrical complication — just the opposite of the make-and-break. Most engineers considered the types produced equal power, but some manufacturers found that their engines turned an extra 100 rpm when equipped with a jump-spark system and sometimes rated such engines one horsepower higher than similar make-and-break models. The timer of a jump-spark engine was driven by a shaft that meshed with a bevel gear on the crankshaft. This was true on a one-lunger, and it was true on Buffalo's unusual four-cylinder. On a single, the timer itself often consisted of a brass case in which a revolving spool was mounted. The spool was composed of fiber insulation material, except for a brass segment that came in contact with a terminal mounted on the case. As the spool revolved, contact was made and broken. Such timers, with more segments on the spool, could suffice for multi-cylinder engines. Other types of timers existed, too, some rather closely resembling a distributor.

Frank Hawboldt designed a jump-spark timer that owed something to the make-and-break in appearance. It is illustrated here. This timer was based upon a plunger that rode on a cam lobe on the end of the crankshaft between the flywheel and the base. On the upper end of the plunger was a sleeve containing a spring-loaded ball. Above that, suspended on an arm, was an ad-

Jump-spark timer by Bemus.

Early distributor and "timer."

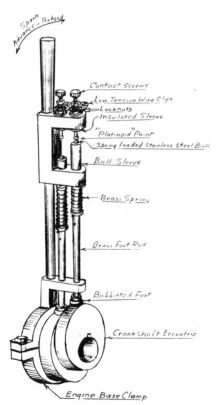

Spark
Advance – Retard

Contact Screws
Low Tension Wire Clips
Locknuts
Insulated Sleeve
"Platinoid" Point
Spring loaded Stainless Steel Ball
Ball Sleeve

Brass Spring

Brass Foot Rod

Babbitted Foot

Crankshaft Eccentric

Engine Base Clamp

Sketch of Lever type timer
for Jump Spark two-cycle
engine invented and first
used by F C Hawboldt, Chester,
Nova Scotia, about 1905

G. H. Hawboldt
1981

justable contact point made of weather-resistant alloy. The plunger was moved upward by the cam, making contact with the point and sending a low-tension current to the coil, where it was stepped up to fire the spark plug. The plungers were mounted together with a vertical lever that, when moved from side to side, caused the plunger to vary its position on the cam's lobe and advanced or retarded the spark. The system was practical for one- and two-cylinder engines — it would have been too cumbersome for anything more complex — and was imitated by a number of other makers.

At the heart of the jump-spark ignition system was its high-tension coil. While a low-tension coil was composed of an iron core and a primary winding of wire, the high-tension coil had a second winding of fine wire. Making intermittent contact between the primary winding and the battery or low-tension magneto, thus providing a constant flow of sparks from the coil, was an interrupter or vibrator. It operated electromagnetically, several thousands of times each minute, and accounted for the hum or buzz that emanated from the coil's varnished wooden box.

Although engine makers were quick to admit that the jump-spark system was more efficient than the make-and-break, few denied that its sensitivity to moisture could be a shortcoming. Wires had to be insulated well and kept dry. That applied to the coil and ignition switch as well. Keeping everything dry in an open fishing boat, however, was not simple, so most fishermen seemed to opt for the make-and-break. Those who didn't sometimes took advantage of the jump-spark's sensitivity to moisture to suit themselves. A squirt of tobacco juice, aimed squarely at the spark plug, was a handy way to stop a jump-spark engine.

Whether jump-spark or make-and-break, the two-cycle engine was capable

Early spark coils: low tension on left, high tension on right.

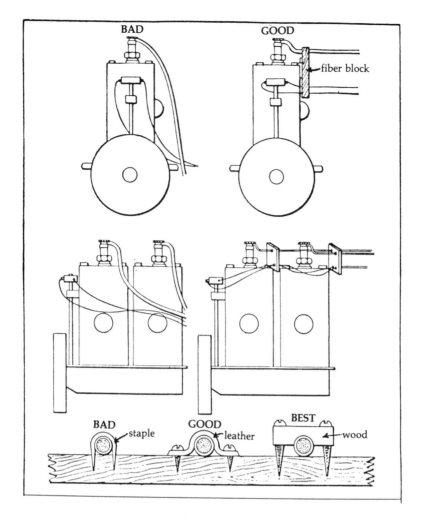

Good and bad methods of ignition wiring.

of being reversed "on the switch." The ignition was controlled by a push button or knife switch. By opening the circuit, retarding the spark, and then closing the circuit again just as the piston came up on compression, direction could be changed. Old fishermen with intimate knowledge of their engine could perform this maneuver pretty much infallibly and generally believed that a man who couldn't manage a boat without a reverse gear didn't know much about boats. A novice trying to reverse on the switch was likely either to stall the engine or to keep it running forward, sometimes with disastrous consequences to docks, his own boat, or somebody else's.

Engines fitted to run on both gasoline and kerosene operated on the same ignition system, no matter which fuel was used. Engines intended only for oil use commonly were fitted with hot bulbs or hot tubes heated either by blowtorches, just as they had been in the very earliest days, or electrically. By the end of World War I, compression ignition engines similar to Diesel's were also available. They were especially popular in Scandinavia and, as discussed in Chapter 1, in the Pacific Northwest.

LUBRICATION

Two-cycle marine engines were equipped with several types of lubrication systems. These included drip-feed gravity oilers, splash lubrication, mechanically driven oil pumps, compression oilers that used the pressure in the base to pump oil, and — ultimately — the mixing of oil with the gasoline itself. Often a combination of systems was used. Early Loziers, to cite but one example, had the crank and rod bearings lubricated by crankcase oil splashed about by a cup, or "dasher," attached to the lower end of the connecting rod. The cylinder walls and piston were lubricated from a cylinder-mounted oiler operated by crankcase suction via a one-way check valve.

The name of the first person to conceive of simply mixing oil and gas together, and the date he did so, have been forgotten, but some manufacturers were recommending this method by 1910. In 1913, the Standard Company of Torrington, Connecticut — makers of the Eagle — claimed they had been the first to recommend lubricating by an oil/gas mixture. "This method was thoroughly tested in our laboratory and found to be exceptionally efficient and a most successful method of lubricating all of the interior working parts of an engine," said the company. "Mix the oil thoroughly with the gasoline before placing it in the tank. You will get better results in handling it this way and you will avoid the possibility of forgetting to put the oil in the tank as has been known to happen."

The recommended oil/gas ratio varied somewhat, depending on who was doing the recommending, but one pint to four gallons generally was suggested. Leaner mixtures were advised — sometimes as little as one pint to seven gallons — if excessive carbon seemed to be forming. Whatever the ratio used, most manufacturers believed they were playing it safe to continue equipping engines with drip-feed oilers too. Some did this for a decade after they began to suggest mixing oil and gas together, but eventually the drip-feed oiler was removed.

Among the reasons the oil/gas method of lubrication found so much favor were its inherent simplicity and the fact that force-fed or gravity lubricators were subject to occasional failure. "In three years only two Elbridge Engines have been fitted with mechanical oiling devices," said an Elbridge catalog in 1911. "Costly force-feed oiling machines were put on these two engines at the expense of the purchaser. The only complaints of imperfect lubrication we have had in three years came from those two engines." Besides possible breakdown of a mechanical system, the dainty brass oil lines that snaked their way artfully here and there about the engine were subject to blockage by carbon deposits. Owners were cautioned to use only oil intended expressly for gas engines to reduce this and related problems.

"If at any time it is found that the piston rings are rusted fast in the grooves, it is a sign that improper oil has been employed," cautioned *The Rudder*. "An unusual deposit of carbon in the cylinder or in the exhaust passages is also an indication of an imperfect oil, and this result may be traced to oils of too great specific gravity or to those which have a proportion of animal oils."

Acadia make-and-break with drip-feed lubricator.

Even an engine run on the day's best motor oil, however, was subject to carbon buildup. Fisherman Gerald Smith remembers that the one-lungers would "carbon up something fierce, especially in back of the rings." Smith recalls that he was cautioned against mixing oil with gas because the situation could thus be worsened. About once a year, he said, he would decarbonize his engine, scraping carbon off the piston top, out of the ports, and out of the ring grooves.

An engine lubricated by a mixture of oil and gasoline was assured of good lubrication, since globules of oil were deposited instantly throughout. This could not be said about gravity-feed oilers, because the oil took some time to make its way throughout the engine. Gravity oilers were simple to operate, but they did require an owner's attention. He had to make sure they had oil in them, and he had to make sure it was flowing. He had to control the flow rate by turning a knurled brass knob at the lubricator's bottom. Five to 15 drops per minute was usually deemed sufficient, and too much oil was liable to foul the spark plug or the igniter. Oil introduced by this method entered the cylinder, passed down the connecting rod, and lubricated the wrist pin through specially bored holes. The oil collected in the base where it was scooped up to lubricate the crankpin.

The rod bearing was the most difficult part of the two-cycle engine to lubricate effectively, and it was the bearing subjected to the heaviest loads. A removable plate was fitted to the base or cylinder so that the bearing could be reached for adjustment without dismantling the engine. Sometimes, a felt wick pad was inserted in the bearing to soak up and retain oil. Two or three thicknesses of flat lamp wick might also be used. Said one engineer who examined a bearing fitted with a wick: "I have frequently examined them after they have been in use and can always squeeze three or four drops of oil out of them."

Pressure-fed lubrication systems were as complex as the drip feed or oil/gas mixture was simple. Gray's Model S lubricator — developed in 1908 — depended upon a belt-driven, valveless pump that raised oil to a sight-feed bowl and then sent it through oil lines to the bearings and cylinder walls. The pump contained four times the amount of oil that a sight-feed reservoir could hold, which saved the operator from having to fill up while his boat was leaping about. This oil pump operated at several hundred pounds pressure, according to Gray, sufficient to remove any obstruction from the feed lines.

Gray emphasized the pressure developed by its oil pump partly as a way of suggesting that manufacturers who relied on base compression to force-feed oil were doing the public no favor. Ferro, however, used just such a system and believed the five pounds pressure developed to be more than enough. Ferro's was an elaborate lubricator, perhaps as elaborate as any fitted to a one-lunger. It relied on an airtight oil reservoir cast integral with the motor's base. The same action of the piston that compressed the air/gas mixture in the base forced open a check valve and sent oil out of the reservoir to a sight-feed distributor. From there, the oil passed out through adjustable needle valves to four oil lines on a single-cylinder, six on a double-cylinder, or eight on a three-cylinder engine.

On the one-lunger, a tube carried oil directly to the cylinder wall opposite the hollow wrist pin and the oil grooves machined into the piston. Some of the oil passed through holes drilled into the wrist pin and then flowed down the rod to the crankpin. Two other lines led directly to the main bearings. Holes drilled in the crank carried oil to the rod bearing. These same oil lines served the forward thrust bearing and the water pump eccentric and rear thrust bearing. The fourth oil line fed directly into the crankcase inlet port, entering the engine along with the gasoline.

Such lubrication systems as Ferro's, the Gray Model S, and others like them, elaborate as they were, did not achieve the simplicity extolled by two-cycle disciples. Certainly, such systems were more complex than necessary in engines that seldom turned more than 500 or 600 rpm, and it is hard to imagine a workboat skipper viewing all those brass feed lines and the pump itself with anything but suspicion. It is likely that by 1915 most of the two-cycle one-lungers had been pretty well standardized with gas/oil mixture lubrication and crankshaft grease fittings. Larger models did often retain gravity lubricators as well. These always were installed on motors to be operated on kerosene, for it was not practical to add lubricating oil to this fuel. Some manufacturers, of course, continued elaborate lubrication systems on one-lungers for years. On its 1923 models, Lathrop noted that lubrication was achieved by three separate systems: gravity-feed cups, a pressure-feed pump, and an oil/gas mixture. So Lathrop had covered all possibilities.

MUFFLERS

By the turn of the century, most engineers were aware that it was desirable to let the two-cycle's burnt exhaust gases escape with as little resistance as possible. Nobody wanted his hot exhaust gases to go back into the cylinder, where they could mix with the fresh charge, cause pre-ignition, and spoil things. Insofar as mufflers were concerned, the problem was not merely to reduce noise but to do so without increasing back pressure and cutting down the engine's performance.

"Of mufflers and silencers there are great numbers," wrote Victor Page in his 1920 *Motor Boats and Boat Motors*, "and many of them are very inefficient." He proceeded to note that the one-lunger was more difficult to silence than a multi, since the latter was "partly quieted by the explosions themselves deadening one another as they issue rapidly from the various cylinders."

Some companies paid considerable attention to silencing their engines, while others did almost nothing, depending in some degree on who was expected to purchase the motor. A one-lunger destined for gentle afternoons in a graceful launch was more likely to be muffled than one installed in a fishing craft.

"The idea we used was just to put the exhaust pipe over the side through the washboard," remembered Acadia's Stan Forbes. "An underwater exhaust cut down on power, and the fishermen never thought much about mufflers, anyway."

The typical muffler, often referred to as an "air muffler," was composed of iron tubing with internal baffle plates. The plates were drilled to permit the flow of gases through each section. Such mufflers were reasonably efficient, and it was a rare company that managed to improve upon them. One that did was Ferro, which as early as 1910 fitted its muffler with cone-shaped expansion chambers. The sudden expansion of burnt gases within the muffler's cones removed "the heat, reducing the back pressure below that of the atmosphere, and allowing the gas to escape without noise and back pressure." The cones in such mufflers yielded a certain degree of suction and were early efforts toward the tuned exhaust system. Such silencers were often called "ejector mufflers."

Sometimes, cooling water was directed into a muffler with water passages constructed within the muffler's cylindrical walls. The volume of water served literally to dampen exhaust noise. It also cooled the muffler and reduced the chances that it would scorch adjacent woodwork. The Syracuse Gas Engine Works patented an exhaust system in which the entire volume of cooling water

Sectional view of standard muffler showing interior arrangement. (Ferro Engines)

came into contact with the exhaust gases in a specially designed manifold. "Thus," said Syracuse, "at one stroke we obtain a clean mixture, a cool exhaust bonnet, and reduce back pressure. The hot exhaust passing out strikes the plate with the cold incoming water and is quickly condensed. Back pressure is thus reduced and power is SAVED while being INCREASED, because of the better scavenging of the cylinders and a cleaner mixture being allowed for each charge."

The greatest sound deadener of all lay just below the waterline, and underwater exhausts occasionally were devised. Most also retained a small muffler to break up the violent pulses of exhaust gas before they entered the water. Such exhaust systems had to be thought out carefully and the exhaust outlet located precisely. Otherwise the same pressure that displaced the water at the outlet would create back pressure in the engine and reduce power. Authors of books and brochures on gas engines, taking nothing for granted, pointed out that all such underwater exhausts should face backward! They also suggested fitting a petcock to the system. This could be opened to reduce pressure and make starting easier. Closed after the engine had stopped, it prevented water from being drawn back into the cylinder by any vacuum remaining within the pipe.

As far as the fisherman was concerned, however, mufflers, expansion chambers, and underwater exhaust systems played no significant role. The greatest concession a fisherman might make to muffling his one-lunger was a piece of high-temperature hose fitted over a pipe screwed into the exhaust manifold. Five or six feet in length, the hose itself made a surprisingly efficient muffler. Otherwise, the fisherman's one-lunger ran with a straight-through pipe, and the healthy bark of those old two-cycles echoed in harbors all around the coasts.

By World War I, the two-cycle marine engine had reached a state of development that characterized it for decades to come. Occasional experiments were made with induction systems — including reed and rotary valves — and other components, but the basic two-cycle one-lunger was never substantially improved upon. An engine built by Acadia in 1955 still had parts interchangeable with one built in 1915.

That great advances in design or materials were limited proved to be no bad thing. Here was an engine whose operation depended upon very basic mechanical and electrical principles that were understandable, in a practical way, to thousands of people. Whatever drawbacks the two-cycle engine may have had, it could be said that never again would a boatman be shipmates with its kind of approachability, simplicity, and — can one deny it? — personality.

These machines were very simple as engines go and possessed amazing efficiency and power for their size. They turned a big propeller at slow speed, and although rated at only five to ten horsepower, the "horses" were all Percherons . . . ."

John F. Leavitt, *Wake of the Coasters*

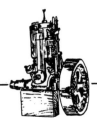

4

How They Built Them

ALTHOUGH TWO-CYCLE MARINE ENGINES were sometimes built in barns, tiny machine shops, or foundries employing a handful of people, they were also built by large companies with fine brick factories. Here were found the lathes, boring mills, and other machines operated by the usual labyrinth of spinning overhead driveshafts and leather belts. Companies with modern facilities liked to portray them in their catalogs. Invariably the drawings showed rows of sleek structures with smokestacks belching prosperously. It apparently bothered nobody that such renderings tended to make everybody's factory look the same. The main objective was to show customers how modern and up to date you were, that your engine was *not* built in a barn.

In the factories, work was hard, noisy, and, especially in the foundry, dirty. It was a bit dangerous, too, since the many machines stood in wait for the unwary. Fatal accidents happened on occasion, and more than one surviving old-timer is missing a finger or two, or has a faded scar to show for his years in the shops. The level of workmanship remained quite high, however, and demanding foremen kept a tight rein.

A work day often began at 7:30 in the morning and continued until 5:00 or

Machine shop, Fay & Bowen Engine Company. This was the company's shop at Auburn, New York, about 1903. (Courtesy Geneva Historical Society)

Machine shop, Torrey Roller Bushing Works, Bath, Maine. (Courtesy Maine Maritime Museum)

Torrey's foundry. (Courtesy Maine Maritime Museum)

Block and head castings, Erd Motors Corporation, Saginaw, Michigan, 1925.

5:30 in the evening. Sometimes the men were paid by the hour; sometimes they were paid on a piecework basis. On Saturday, they worked a half day. Skilled workers were for the most part made, not born. A man might begin his career sweeping up the machine shop floor or rolling wheelbarrows full of castings from one part of the plant to another. Gradually, he worked at various jobs until he understood the overall picture. Then he might settle down in the foundry or some other area that appealed to him. At Acadia, one man made pistons for 50 years.

The ultimate quality of the one-lunger depended to a great degree on the skill of those who built its component parts. If the design of a two-cycle engine placed demands upon the engineer, its casting was an intricate business that demanded great understanding on the part of the foundrymen. The fact that many of these engines did not have a detachable head — hence the nickname "headless" or "domehead" — only complicated the process. Engines were cast using a set of wooden patterns, one for the outside of the cylinder and two more, called core boxes, for the combustion chamber and water jacket. Such patterns were vastly complex. They required the patternmaker to know his craft thoroughly, and the patterns and cores, when completed, were wonderful to behold. The wood was joined and shaped to minute tolerances.

The various parts of the sand mold and the cores had to be positioned exactly before the pouring process was begun. Otherwise, the casting would have a thin spot here and a thick part there. This could, and did, lead to warpage and breakage. Misalignment of the core was sometimes referred to as "floating." It could result in a portion of the cylinder wall or water jacket being only $\frac{1}{16}$ inch thick instead of $\frac{3}{16}$ inch.

CYLINDERS

The outside of the cylinder was cast in molding sand. At Acadia, most of this sand came from Albany, New York, and was called Albany molding sand. It had enough clay content so that it could be molded by being mixed with water. There was a sort of artistry to this mixing process, and the old molders developed an instinct for just how much moisture the sand needed. They seldom used instruments of any kind. They just mixed the sand with water on the molding floor, shoveling it this way and that. If it looked too dry, a splash of water was added from a bucket. If it looked too wet, more sand was added. After a while, the molder would bend down, grab a handful of sand, and squeeze it. He knew by its feel when it was just right for molding.

When the sand was ready, the pattern was placed in the molding box or flask and the sand shoveled in on top of it. Then it was pounded all around the pattern with a hand rammer. If it was not rammed tightly enough, molten iron would push the mold out of shape and cause a defect in the casting. If the sand was packed too tightly when the molten iron ran in, the resultant gases couldn't escape. "When that happened," said Stan Forbes, "why, the thing would kind of blow up and ruin the casting." Forbes spent 38 years at Acadia, most of them as supervising engineer.

Molding room, Lunenburg Foundry.

Packing sand in mold, Lunenburg Foundry.

Stan Forbes and milling machine at Acadia, June 1981.

After the flask was properly prepared, the pattern was removed and the various cores were set in place. The sand used for the cores was of a special type, although, especially in the early years of engine manufacture, natural beach or river sand was often used. The sand had to be clear, without mud or salt. At Acadia, such sand was delivered by schooners that made their way up the La Have River and unloaded their supply directly through hatches in a shed near the edge of the wharf behind the foundry. The sand came from local beaches and sandbanks formed above sea level, which had been cleared of salt and impurities by rain.

The sand was mixed with an oil called core oil and then packed into the core boxes. After the sand had been fully pounded around the pattern, the core box was placed on a flat metal sheet, and the sand mold that had been created was removed. Still on its iron sheet, the mold was placed on a carriage and rolled along a track into a large oven. The oven door was shut and the temperature raised to some 250 or 300 degrees Fahrenheit. After that, the oven was turned off, and the heat retained in the brick lining finished the baking process overnight. Cores thus baked could be handled without breaking if care were taken. "By then, the sand had about the same consistency as a baked cookie," said Stan Forbes. Smaller cores were handled in a similar fashion, although they were baked in a much smaller oven on shelves that revolved much like a lazy Susan. The small pieces required only about a half hour's baking time.

With the sand cores prepared, the molder placed them precisely into the flask holding the fragile, green-sand cylinder mold. After the cores were aligned, the second half of the flask, containing the other half of the cylinder, was positioned and clamped. Care had to be taken lest the green sand in the top half be disturbed unduly or even fall out. The joined flasks were now ready to receive the molten iron that had been melted in a cupola and then drained into a crucible or ladle. This iron was poured in through an opening in the top of the flask. The metal used for cylinders and bases was close-grained iron, always referred to as "gray iron." Often, manufacturers made such iron from scrap, including discarded flywheels and other pieces. The iron frequently was used straight, with nothing else added. Sometimes, however, one percent of nickel or chromium — purchased in either shot or bar form — was used to make the iron more dense and presumably longer wearing.

After the casting was poured and the iron had cooled, the completed part was removed from the sand mold and placed in a tumbling barrel or mill, a belt-driven iron cylinder. Castings were loaded rather tightly into the tumbler, together with an appropriate number of metal pellets, often referred to as "bugs." Tumblers revolved with their burden of fresh castings, making a tremendous racket. The movement of the castings and the friction of the "bugs" loosened remaining sand from within the casting and cleaned the ex-

Tumbling mill, Lunenburg Foundry.

terior as well. If any sand were left in the water jacket, it might expand, once the engine was put into service, and crack the casting. It was easier to clean a cylinder casting for a detachable-head engine. Here, passages could be cleared physically, if necessary, with a rod or wire. Of course, such engines did have an extra joint that required a gasket, usually one made of copper-clad asbestos. Makers of "headless" engines liked to point out that a head gasket was always prone to failure.

Fresh castings were usually placed in rows in the foundry yard for two or three months to cure. Heated by the sun during the day, cooled at night, stresses within the iron were relieved. The process accounted for the junkyard impression a casual visitor to the foundry would receive, for the curing castings soon developed a coating of rust. It was not until World War II that green castings were frozen with liquid oxygen in an immediate stress-relieving process.

At Acadia, Stan Forbes said the newly cast cylinders were put in racks in a shed. "We usually did 30 castings in a lot," he remembers. "We had 30 in the racks, 30 in the machine shop, and 30 on the shelf waiting to be taken in for assembly." Often, castings were tested hydraulically. The water jacket was sealed and water injected to 100 pounds pressure. If there was any weeping into the cylinder, the casting was porous and was set aside.

FINISHING THE CYLINDER

Completed cylinders were usually bored on a vertical boring mill, the cylinder held in a jig while the boring tool did its work. Once the cylinder was roughly bored to size, it was ground to exact diameter and then honed to glasslike smoothness. Sometimes, while the cylinder was being bored, its bottom — where it would be joined to the base — was simultaneously finished, all the surfaces made square and true. Mounting holes were drilled, usually in a jig, so that various cylinders and bases would be interchangeable. Builders of high-quality engines always took pride in their cylinder boring methods, and, if they felt they did something particularly special, they made sure prospective buyers knew about it. Here is what Northwestern Marine Motors said of its cylinder-boring operation in 1911:

> In the ordinary cheap grades of motors, the work is hurried along in every particular. The tools are set to cut deep, so as to finish boring the cylinder in two or three cuts. This makes an unsatisfactory working engine, because the tools will spring, causing the cylinder to vary in roundness. The cutting often heats the walls of the cylinder, causing it to expand where the iron is thin and remain normal where the ports and other connections are, between the inside and outside of the water jacket. This causes the iron to give away from the boring tools and necessarily causes high spots in the cylinder when bored, giving a poor compression and uneven wear of the piston rings.
>
> To overcome these difficulties, the cylinders are bored too large for the pistons, consequently giving a poor fit and a noisy running engine. In the Northwestern engine, at least four cuts are taken with the boring bars and one with the reamer and polishing grinders . . . .

Cylinder boring machine with cylinder in place, Lunenburg Foundry.

Finishing the cylinder's exterior required a heavy, paste-like filler that was applied and sanded smooth. Then the engine was painted with glossy enamel. How much care went into finishing depended to a great extent on how expensive the engine was. Cheaper motors might receive a quick filler coat and a single coat of color. Most engines received two to eight coats of paint, and this, combined with their bright brasswork, engraved nameplates, and gleaming bolts, yielded some handsome machines. At Acadia, a heat-resistant varnish was used in the finishing process. It was followed by a coat of gray filler that was sanded smooth before the final coat of gray enamel was applied. Sometimes a finish coat of varnish was added. At Lozier, each engine received three coats of oven-baked black enamel. For a slight extra charge, the cylinder head was nickel-plated, producing a "special finish engine" that was "an ornament to any boat."

Gray, black, red, and blue were commonly used colors. Brown and green were also popular, although the latter was seldom seen in Newfoundland, where it was considered to bring bad luck to a boat. French fishermen along the St. Lawrence discovered that engines painted entirely with red lead would change shades if running too hot. This signaled that it was time to investigate the water pump. For this reason, no matter what color a manufacturer painted his engines, the St. Lawrence fishermen promptly covered them with red lead.

PISTONS AND RODS

Pistons were almost uniformly cast of the same gray iron as the cylinder, ensuring that expansion rates would remain the same as the engine warmed up. This was a more important consideration than reducing the piston's weight by making it of aluminum, although this was tried from time to time, unsuccessfully. Sometimes the experiment was carried out by the manufacturer; sometimes an owner might conduct it. Naval architect John Atkin remembers that his father, William, removed the iron piston from a Lathrop one-lunger and had Lathrop make him a new, lighter piston of aluminum.

"No one figured that the aluminum would expand faster than the iron, and the thing seized up," John Atkin recalls. "Then we remachined the piston and took off so many thousandths that the engine never ran worth a darn. It four-cycled and, for some reason, ran even better in reverse than forward."

Attempts to improve on the basic one-lunger, either with new materials or new design ideas generated much after 1900, seemed mostly doomed to failure.

Pistons usually were fitted with three rings of hardened iron, sometimes four. Usually there were two or three at the top and one at the skirt. There were two basic methods of shaping the ends of the rings to provide a seal. One was to cut the ring at a 45-degree angle. The other was to form a notch or step — called a lap joint — in each end of the ring so that the steps interlocked when the ring was installed. The lap-joint method was superior, since there was less chance of leakage as the ring gradually became worn. Most manufacturers, realizing the importance that the rings played in sealing, paid special attention to their fit and finish. The rings were ground eccentric so that, once installed in the lands, they would form a perfect circle. The outer surface was highly polished to reduce friction. The rings were kept from rotating on the piston — and thus from snagging a port and breaking — by tiny pins. The pins were inserted in specially drilled holes in the ring grooves. A notch in the ring mated with the pin.

Although the practice was not universal, it was not unusual for pistons to have a groove, or "oil gain," machined around their mid-sections, the intent being for this depression to carry added quantities of oil to the wrist pin and cylinder wall. The wrist pin itself was machined from hardened steel and was mounted in phosphor-bronze bushings. Sometimes the wrist pin was held in place by set screws. These could, and occasionally did, work loose, allowing the pin to be displaced and score the cylinder walls. Most builders developed a better way of retaining the pin. Sometimes the wrist pin was secured to the rod so that it could not revolve at all in the piston.

The fitting of the piston and rings to the cylinder was a job that required skill and experience. The goal was to have things as tight as possible without being too tight. Some companies boasted that tolerances of .001 inch were maintained between cylinder and rings, but one must question how accurate such statements were. Stewart Greer, an engine fitter at Acadia, recalls that most of the engines were set up entirely by feel, without even the use of a feeler gauge. "You had to check, especially on the 3 and 4 hp," he said. "If clearance was too great, they wouldn't develop enough power. The fit was less critical

Cylinder and piston patterns, Lunenburg Foundry.

on larger engines. You could pretty well tell when it was right: it was all feel. You did try for little variation from one engine to the next and always tried to match a piston to a cylinder. It was a challenge."

BASE, BEARINGS, AND CRANKSHAFT

The base was cast of the same gray iron as the cylinder and piston. So that the babbitt bearings would have something to hold to when poured, several ⅛-inch-deep holes were drilled into the base — and the cylinder half, too — at a 45-degree angle. Occasionally, the inside of the base was honed and polished to a smooth finish. More often, however, it was given a coat of red lead to seal in any dust remaining after the tumbling process had been completed and to keep any particles out of the rod journal.

Before the bearings were poured, the base and top half of the bearing shell were heated so that the bearing metal would flow and adhere properly. The bearings were poured around a carefully aligned mandrel that was the same diameter as the crankshaft. The mandrel was supported at either end by iron bosses that looked exactly like large, thick washers. These kept the molten babbitt from merely flowing out of the engine onto the workbench or floor. Some, perhaps most, companies babbitted the bottom half of the base and the top half, or cylinder half, separately. At other firms, the cylinder was mounted to the base atop a temporary asbestos gasket and the babbitt was poured in through the grease cup holes. Once cold, excess babbitt was filed off flush using a coarse rasp. Each maker had his particular preference in babbitt material. Some favored metals with a comparatively high content of nickel and tin, while others preferred more lead, nickel, and zinc in the alloy.

As we have seen, provision was made for taking up the wear of the main bearings by inserting shims of thin metal or greased paper strips that could be added or removed as necessary. Eventually the bearings would have to be

Cutaway Acadia engine. On wall at left are washers used to hold crank while bearings are poured.

rebabbitted. To avoid this periodic chore, some engines were designed with replaceable bearings, which usually were made of bronze or brass. Especially during the early years of the gas-engine movement, these were preferred by some over babbitted bearings, particularly when an inexperienced operator was in charge of the engine. "An overheated bearing is quite likely to cause the babbit metal to melt and throw the engine out of commission," said an article in *The Rudder* in 1900. The Ferro company tried to use the best features of both bearing types. The upper half of the Ferro's main bearings were made of bronze castings lined with babbitt. They were fitted with shims to take up wear and, when the babbitt was well worn, were a simple matter to replace. The lower halves were poured directly into the base.

Bearings were reamed, sometimes by hand but more often with the aid of a reaming jig that mounted a heavy-duty electric drill. That some workmen could align and smooth the main bearings without such aids is testimony to their skill and years of experience. Bluing smeared onto the crankshaft was used to reveal any high spots. By the time the bearings were ready, they were uniformly blue. Many companies claimed the tolerances at their engine's main bearings were .001 inch, and those who had witnessed a skilled mechanic doing this work did not doubt such claims.

The bearing was not complete until a shallow groove had been cut in by hand leading out along the bearing surface from the grease cup hole. At the grease cup, the groove was about ⅛ inch deep, but it gradually became shallower, tapering away to nothing before the end of the bearing. The groove allowed grease to be forced out from the cup along the whole bearing.

Before leaving the subject of bearings, it might be worth noting that the idea of using ball bearings probably occurred to more than one engine builder. Few bothered to document their experiments, and the idea did not find its way into production. Frank Hawboldt, Jr., tried ball bearings in one of the motors his father had designed.

"The problem," he says, "was water. These engines ran half the time submerged in salt water. That would wreck a ball bearing. Then too, you had to have a mighty good seal in those bases, and that was difficult to achieve with a ball bearing."

Like aluminum pistons and most other newfangled notions, ball bearings and one-lungers didn't mix.

Most companies seemed to take a special pride in the crankshafts they used and seldom admitted that any competitor outdid them when it came to this vital component. The finest cranks were made from a billet of drop-forged, high-carbon steel able to withstand all the shock and torque to which it would be subjected. Often the forgings were obtained from a large foundry, made to the individual company's specifications. Most small and medium-sized companies found it economically prohibitive to forge their own cranks and had little hope of doing as good a job as a specialist. The best crankshafts were so tough and malleable that their ends could be bent around like a pretzel without breaking.

A certain variety was evident in the making of crankshafts, as in nearly everything else about two-cycle engines. Occasionally, a company intent upon making everything possible in-house would contrive a way to make crankshafts from flat, forged steel plate. This was a laborious process that will be detailed in Chapter 5. However the crank itself was made, it was always highly polished. Often this was done with the aid of wooden polishing tongs. The tongs' jaws were lined with emery cloth. Set around the main journal or crankpin, the tongs were worked back and forth until the journals were gleaming.

Although a one-lunger could be counted upon to vibrate, especially at low speeds, few two-cycle manufacturers bothered to balance the crankshaft in an effort to smooth things out. Those who did were quick to note the fact. "The fitting of the counter weight on the crank shaft is a very important thing," said Gray in 1911, "and it cost the Gray Motor Company many thousands of dollars to find out how to do it right." Gray made its counterweights of malleable gray iron. The weights were machined to fit and attached with iron dowels and rivets.

Northwestern used balance weights, too. "We have experimented with our 4 hp motor," said the company, "and by building it exactly the same in every particular but leaving out the balance weights, it makes a difference of one horsepower."

Multiple-cylinder engines, with the crankpins set at angles of 180 degrees on a twin or 120 degrees on a triple, were inherently smoother running than one-lungers. Even Northwestern did not balance the cranks of its multis.

Because forward thrust was exerted against the crank as the boat moved ahead, and reverse thrust when going backward, thrust bearings were installed front and rear on the crankshaft. These were often formed of a sandwich made of two outer solid discs and an inner disc that carried many ball bearings. This was counted as the most efficient way to relieve thrust strain and lengthen engine life. Such ball thrust bearings were not used by everyone, of course. The Knox was equipped with plain bronze thrust washers held against the end of the bearings by spring pressure. "They prevented," said the manufacturer, "any possible chance of gas escaping around the end bearing after they begin to show wear, and, as everyone knows, any bearing will show wear in time, and sufficient wear to allow gas to escape."

THE CONNECTING ROD

Connecting rods were made of forged steel or forged bronze. Those who used bronze were likely to suggest that their rod would not crystallize under repeated shocks, and those who used steel claimed the same thing. As long as the alloy was good, both metals proved themselves. Most connecting rods were forged in what today would be called an I-beam configuration. Some companies called this a truss shape or a double-T. Others, thinking of weight savings or improved lubrication of the crankpin, used a hollow, tubular rod. I-beam rods were often drilled their entire length to carry oil to the lower rod bearing. The crankpin bearing was usually secured by two bolts, a practice that is still common. Some engines had a bolt on one side of the rod's lower end and a hinge on the other. This saved on cost and meant there was only one bolt to unfasten when the bearing needed adjustment. Shims were provided so that wear could be taken up after the side cover on the engine had been removed.

FLYWHEELS

The flywheel was cast of iron and fitted to the crankshaft's tapered front by means of a steel key engaging both a groove in the shaft and another in the flywheel. Often, a large nut secured the assembly, and sometimes this nut was covered by a brass sphere so a pants cuff could not become caught in the nut itself. The flywheel key could usually be counted on to corrode after a number of years, and this often made removal of the flywheel a frustrating business.

The flywheel's purpose was to turn the engine's abrupt, individual power impulses into smooth, continuous motion. Some engines had balanced flywheels; others did not.

Said the Peerless Marine Engine Company of Detroit: "Great care is taken in balancing the wheel, before the wheel is placed on the engine, as we claim to have the smoothest running and the freest from vibration of any engine manufactured. A man can enjoy a long run with one of our engines, and not have his nerves addled like he would with one of these cheap, noisy, rattling engines."

Although not all manufacturers took the time to balance the flywheel, all realized the necessity of concentrating weight in the wheel's rim. In order to lighten the center portion, many companies removed metal in sections. Each builder had his own preferred pattern. Some put in four large, round holes; others cut out large triangular sections. Some cast the flywheel with a center section resembling the spokes of a wheel. The Atlantic engine's flywheel had four heart-shaped cutouts in the disc. Often there were no cutouts at all. The solid flywheel was easiest to cast and the one most likely to be balanced. The weights were attached along the inside of the rim. Flywheels were painted the same color as the rest of the engine, but sometimes the rim was painted a contrasting shade or coated with aluminum paint to prevent rust.

Installed in the flywheel was the starting handle. This usually took the form of a small, spring-loaded brass dowel or knob. The spring had to operate smoothly and positively or else the handle would not retract and could easily break a wrist as the motor began turning. Palmer eventually stopped making such devices because of the injuries they caused. Despite the danger, fishermen often removed the spring-loaded handles and replaced them with a metal rod or thick wooden dowel that, of course, remained protruding after the engine was running. Generally, engines so fitted were intended to be started with one's foot. Occasional broken shins were the result.

Pleasure boats were sometimes fitted with automobile-like cranks mounted on a pedestal and connected by chain to the crankshaft. With such an arrangement, a man did not have to kneel awkwardly before his engine, rather like a suppliant at an altar. Some flywheels, such as Lathrop's, had deep serrations cast into the inside of the rim. These provided a fingerhold, and a man with reasonably strong hands could pull the engine smartly back against compression and thus get it going.

Being entirely exposed and so large, the flywheel probably received more attention from owners than it deserved. It was likely to be the first part kicked when a motor misbehaved or patted affectionately after the machine had rendered a day's faithful service. Here is a poem about flywheels that appeared in *The Rudder*:

> When a flywheel doesn't fly
> It is apt to be some shy.
> If it doesn't turn the shaft
> How you goin' to run the craft?

Naval architect Fenwick Williams remembers a Marblehead yachtsman who firmly believed that as long as he painted his flywheel with a fresh coat of red paint each season, the motor would behave. "It worked, too," said Williams.

IGNITERS

Intricate in design, the make-and-break igniter was also difficult to fabricate. Most parts had to be machined, and a good machinist needed six or seven minutes to produce, for example, a movable electrode. He would produce them, and other pieces, by the hundreds while another man made the stationary electrode of mild steel and another cast the igniter base of iron. Sometimes, an electrode was bought from a supplier, together with the mica washers used for insulation, but most companies made their igniters in the factory.

Ferro made its igniter out of hardened steel with platinum points. Torrey Roller Bushing Works made the Kennebec's igniter base out of bronze and used nickel alloy for the points. Opinions differed on the best metal for the igniter base. Camden Anchor-Rockland Machine, makers of the Knox, believed that bronze was not as good as iron for the igniter base. The company said bronze tended to crystallize under the intense heat to which the igniter was subjected.

Whatever the electrode base and points were made of, the movable electrode had to be lapped into the base like a valve. Otherwise, compression would be lost. The electrode's seat was subject to wear and had to be resurfaced, on occasion, just like a valve seat. Sometimes a maker claimed to have found a superior way of fitting the movable electrode. Said one: "Instead of a beveled seat between the igniter plug and the movable electrode, we have found a flat seat far superior in preventing leakage around this electrode after the bearing has commenced to wear."

Of the early days of the two-cycle engine, Frank Hawboldt, Jr., recalls that "in the old make-and-break, the main thing to know was the igniter. The movable electrode used to wear out its seat and the gas could go through it and the engine would lose compression." Hawboldt once experimented with setting the electrode in a mounting of small needle bearings. This worked no better than his ball-bearing crankshaft, since rust soon rendered the needle bearings inoperative.

The low-tension coil that fired these igniters was a simple device. It was made on a wooden frame into which a soft iron core was inserted. An insulated covering over the iron was made of a cardboard tube. Over the tube went the winding of fine wire that constituted the primary winding. The winding itself was done on machines designed specially for the purpose, and the ends of the wound wire were attached to the coil's two electrodes. When the winding was completed, the coil was dipped into a bucket of hot tar. The coating thus formed insulated the whole coil. The coil was then complete.

Low-tension coil by ABCO Acadia. Frame is at right, with toilet-paper cardboard roll separating iron core from winding to be applied.

Coil winder, ABCO Acadia.

ASSEMBLY AND TESTING

Engine fitters, those mechanics entrusted with the final assembly of the motors, worked at their own benches, collecting the bits and pieces they needed from a company stockpile. To an onlooker, it might appear that the fitters went about their job rather casually, but these men had so much skill and experience that they knew, as if instinctively, exactly what they were doing. There was little they couldn't do with a few seemingly casual taps of a mallet or quick adjustments with a wrench. They worked methodically, adding each piece and adjusting it as necessary. When they installed the crank, rod, and

Fay & Bowen — motors are being limbered up prior to testing. Overhead belts drive the flywheel. (Courtesy Geneva Historical Society)

piston and added the top half of the engine cylinder, they checked to see how much force was needed to revolve the crank. If the rod and piston more or less dropped down by themselves, with little effort, the fitter sensed he had things just about right. Pistons usually were set up to have about .002-inch clearance per inch of bore.

Adjusting the igniter mechanism was a demanding task. The various rods, springs, and levers had to interact correctly or the engine would not be timed correctly. "Usually," said Acadia's Stewart Greer, "I would try to get things so that when the flywheel hit 9 o'clock, the tripper would begin to make. At about 11 o'clock, it would break and the spark would jump." His practiced fingers seemed to know just what little screw or bolt in the system needed to be turned, and how much, to achieve this goal.

It took the engine fitter about 2½ to 2¾ hours to complete a one-lunger. When he was done, when he had tightened the last bolt and oil fitting, the engine was ready to go to the test house. Although he would begin work at once on another engine, the fitter usually kept an eye on the test house to see how his most recent product fared. Usually the engine emitted a satisfying cloud of blue exhaust smoke, but sometimes an ominous silence prevailed.

Carl Tumblin spent two years testing engines at the Lunenburg Foundry.

Standing in his yard, surrounded by old engine parts, he told me about his experiences. In a part of Nova Scotia where many houses are brightly painted, Tumblin's was the brightest of all. It was painted orange and yellow. "The biggest problem in the test house," he remembered, "was that the engine might seize up because it was new and pretty tight. I used to double the amount of oil going into the engine." It was rare, Tumblin said, that he had to return an engine to the workshop. If an engine seemed particularly tight, he might add a shim to a bearing, and, after running the engine for two hours, a shim might be removed. In either case, a hose was kept trained on the main bearings to keep them cool.

In the test house at Lunenburg Foundry — as at Acadia and many other companies — new engines were hooked up to shafts that drove large, paddle-like wooden fans. The diameter of the fans could be varied by bolting on blade sections to provide the degree of resistance for whatever horsepower engine was being run. A chart on the wall specified what size fan was to be used with each engine. Sometimes the engine's new owner came to the factory on the day the engine was being tested so that he could verify that his engine produced the stated number of horsepower.

Such testing methods were common at companies turning out from 500 to 2,000 engines per year. Other methods were used, however, especially by large manufacturers, which often had elaborate facilities capable of testing many engines at the same time. At such factories, motors might be "limbered up" by attaching the flywheel to a belt driven off a stationary power source. Engines were kept turning like this, well lubricated, for as long as 18 hours, sometimes several days.

After the bearings had been worn in, the motor was ready to be started and

Test house, Lunenburg Foundry.

run on its own. Often, performance was checked on a brake. At Ferro, where they were tested by the dozen, engines were hooked up by belt to dynamos. The voltage and amperage that each produced was checked to see that it fell within acceptable levels. The engine was then put on a brake for horsepower tests. It was not until an engine had passed these tests that it was taken to the paint shop and given several coats of oven-baked gray enamel.

At Weco, after the engine had been run in on a factory belt, it was mounted on a test bed identical to one in a company launch. The bed contained a driveshaft and propeller that ran in a large concrete tank "filled with water of sufficient depth to give exactly the same resistance as met with in a boat. Here the engine must run by its own power until it develops the required number of revolutions per minute without undue heating." After that, the motor was run another eight hours. A data card on each engine was filled out, signed by the general superintendent, and kept on file.

In the fiercely competitive world of early boat motors, nobody wanted to let a lemon leave his factory.

How accurate were the horsepower figures stated by the various manufacturers of two-cycle engines? Because several different measuring systems were used, it is hard to say more than that the engines, in general, must have been underrated. Comparing the usable power developed by a 4 hp one-lunger and a modern 4 hp outboard would be like comparing a Clydesdale horse and a Shetland pony.

In 1952, the Canadian Research Council's Division of Mechanical Engineering conducted very carefully controlled tests of a pair of two-cycle engines built by the Lunenburg Foundry. These were a 4 hp and an 8 hp two-cylinder Atlantic. The tests were conducted on a hydraulic dynamometer. The 4 hp proved to develop 5.8 hp at 650 rpm, an output 45 percent greater than the manufacturer's rating. The two-cylinder 8 hp developed 17.2 hp at 950 rpm, 115 percent more than its rating. Although part-throttle operation of the twin was found to be unstable, the engine performed satisfactorily throughout the tests. These included a 100-hour endurance run. After that was complete, the engine was torn down. No significant wear was found, although the top compression ring on each piston was stuck fast in its groove.

Such were the results of the two-cycle's engineering, robust materials, and careful manufacture and testing. The finished product often gave great satisfaction to its owners. Companies delighted in printing testimonials in their brochures. Here are two of them.

Camden Anchor-Rockland Machine Co.   Sept. 26, 1904

Gentlemen:

The 18 ft. dory with 2½ H.P. Knox Engine that I purchased from you three months ago is working very satisfactory and has been running every day since I received her. I would like to know how you succeed in getting so much power in such a small engine. I recently towed a 47 ton schooner loaded with hard wood against a 1½ Knot Tide through the narrows here, and was more than surprised to see the headway she made with such a heavy load and such a small engine. If I were going to buy a hundred engines today I would buy nothing but a Knox. Thanking you for past favors, I remain

Very truly yours,
J.L. Wilmot

Wilmot lived in Lubec, Maine, and the narrows he spoke of are known for tides and eddies so fierce that even modern, powerful fishing boats can have difficulties there.

The following letter was sent to the Lunenburg Foundry in 1949 by a man who had owned an Atlantic engine for rather a longer time than Wilmot had his Knox.

Gentlemen:

I have been using one of your Atlantic Marine Engines for twenty-six years in the fishing boat every working day and in the wintertime setting it up to saw my year's supply of fire wood, and also in cruising back and forth with my business of the village general store. I found the engine has gone, to my own knowledge, one hundred thousand miles during the twenty-six years, with very few new parts and only one vital part, a cylinder.

Arthur Rockwood
Center Cove
Trinity Bay, Newfoundland

DON'TS

Don't crank your head off, look for the cause.
Don't forget to see that the pump is working.
Don't forget to oil the motor frequently.
Don't use anything but gas engine oil in the motors.
Don't use more gasoline than the motor requires.
Don't forget to drain the motor in cold weather.
Don't let the batteries get wet.
Don't let bare wires come in contact with the motor.
Don't let wires run through bilge water.
Don't connect new batteries with old.
Don't run the engine if it is pounding.
Don't forget that ninety per cent of motor failures can be traced to electric troubles; either in the battery, or the coil, or the wiring or the plugs.
Don't read these instructions merely, but study them until you know them: then follow them.

Instruction Book, Atlantic Marine Engines,
Lunenburg Foundry & Engineering Ltd.,
Lunenburg, Nova Scotia, Canada

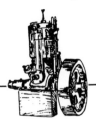

5

Daniel and Charlie and
the Atlantic Marine Engine

OF THE MANY COMPANIES that once existed to build two-cycle marine engines in the United States and Canada, only one still does so today. It is the Lunenburg Foundry, and it has been producing Atlantic engines since 1908. One reason the engine has survived all that time is that it's a good engine. The other reason is that the foundry has always made a variety of products and has never depended solely on sales of the Atlantic for its prosperity. The foundry was begun by three Yarmouthmen — Philo Harris and Charles Patterson, molders, and William Sanders, a patternmaker — and W.T. Lindsay, Lunenburg's town clerk, in December 1891. They planned to make cast-iron stoves, but eventually a vast array of marine-related hardware was turned out. The pattern room is now bursting with precisely made wooden models of gears, portlights, cowl ventilators, windlasses, pumps, winches, and capstans. There are those who have said, only half in jest, that the Lunenburg Foundry could cast a tolerable suit out of iron.

The quartet built the foundry, which they called the Lunenburg Iron Company, on the town's waterfront. Economic activity remains centered here, for Lunenburg's steep streets lead inevitably to the harbor and its shipyards.

Once, a hundred fishing schooners were berthed here. The foundry is only a few minutes' walk from the Smith and Rhuland yards and is within sight of the spot where the most famous schooner of them all was launched in 1921. The *Bluenose* made a lasting impression upon those who saw or sailed her, or vainly raced against her. Eventually she was immortalized under full sail on the Canadian dime. Lunenburg Foundry made much of the vessel's hardware and, in 1935, installed the schooner's two Fairbanks-Morse diesels.

Within a decade of its founding, the Lunenburg Ironworks had become a substantial enterprise, as solid as one of its cast-iron stove tops. It was already larger than many gas-engine companies would ever be. The foundry was housed in a 60-by-100-foot building containing the core room, the mill room with its tumbling barrels, the stove fitting shop, and, above that, the pattern shop. The patterns were stored nearby in the same 30-by-40-foot building that housed the machine shop. There was an 80-by-40-foot warehouse in which the company office was located. There was a building for the storage of molding sand and for the bituminous coal used to power the stationary steam engine and to fuel the blacksmith's forge. There was a drying house where a stove was always kept going during winter months. Pine cut in the abundant Nova Scotia forests was stored here — 5,000 to 6,000 sweet-smelling board feet of it.

One evening in 1898, when the foundry cupola was puffing smoke into the sky above the harbor, three teenagers slipped inside to see what was going on. A two-ton casting was about to be poured, and they watched as a furnaceman used a pointed iron bar to break through the clay bot, or plug, permitting the iron to flow from the cupola into the waiting ladle. A cable was affixed to the ladle so that a crane could hoist it and the pouring could begin. In the ladle, a crust had formed over the iron, and the furnaceman used a crowbar to break through the crust. The ladle was tipped slowly, and the iron flowed through a hole about three inches in diameter. The three boys were watching all this in fascination, when suddenly the crust gave way and the iron spilled out in a shower of sparks, spreading across the floor, scattering the workmen, and sending the boys hopping on top of a pile of molding sand. When things were again under control, they were abruptly sent home.

One of the trio was named Daniel Young. He was 18 years old at the time. Ten years after he and his friends were chased out of the foundry, he designed the two-cycle engine that became known as the Atlantic. For years he was entirely responsible for any changes made to the engine. The next man to step in with more substantial revisions was Dan Young's son.

It would be wrong, perhaps, to say that Dan Young was typical of those who created the marine-engine industry or its relative, the automobile industry. Yet he was not atypical, except for the considerable success he achieved. More is known about him than about most other marine-engine pioneers because the Lunenburg Foundry still exists, his son was always close to him, and, at some point in life, Dan Young began to think that others might one day be interested in the world he had known. When that thought occurred to him, he sat down and, in his own hand, wrote the story of his early years on lined notepaper. He called it *Foundry Experiences: Reminiscences of D.E. Young.*

Like all successful pioneers of the gas-engine movement, Dan Young possessed innate mechanical ability. He could do things with his hands. He fostered his intuitive understanding of machines by reading about them in all the books and journals of the day. He did this, although his schooling took him no further than fourth grade. He had an infinite capacity for hard work and a desire to do things properly. What he lacked in training and schooling, he made up for with effort and common sense, solving problems by trial and error.

As with most of those who pioneered the new industry, Dan Young was born on a farm. The Young farm was in Martins Brook, Nova Scotia, about two miles outside Lunenburg. There Dan Young spent such spare time as he had in his father's woodworking shop and became adept with all sorts of tools. One day, using a jackknife and some special tools he had made, he set about making a particularly ambitious and intricate carving. When he described it later in his diary, it required a page and a half. He had taken a 2¾-inch pine dowel about 10 inches long and turned it into interlocking chain links, spheres in which tiny balls rolled freely about, and a bell, complete with clapper. All the parts were interlocked, all carved out of that single piece of pine.

"I managed," he wrote, "to make the entire carving in one piece without breaking it."

His cousin, Arlington Rafuse, saw the carving and suggested that Dan Young apply for a job where he himself worked, at the Lunenburg Foundry's patternmaking shop. In 1899, one year after he had hopped onto the pile of molding sand to escape the fiery iron, Dan Young started work at the foundry. He went to work at 7:30 in the morning and went home at 6:00 in the evening. On Saturdays, he worked for five hours.

The company's pattern shop was equipped with a circular saw, a jigsaw, a planer, and a grinder. Young was just becoming familiar with these tools after

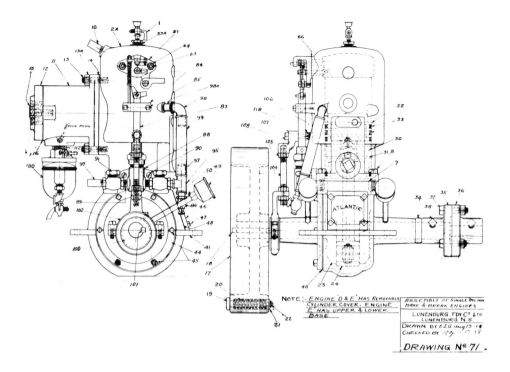

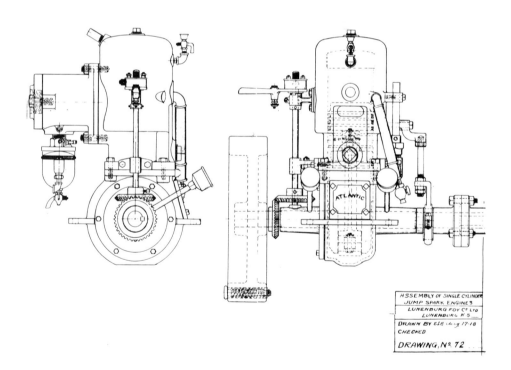

two weeks on the job when a board he was ripping on the circular saw was flung back at him. The board struck him in the face like a spear and he was lucky to suffer only a broken nose. Later in his career, he saw a man killed instantly in a similar accident, and, like all workers in the factories of that era, he became wary of the many spinning belts, shafts, and blades that lay in wait for the inattentive. Once he saw a workmate become entangled in an overhead driveshaft. The man's clothes were stripped from his body, leaving him standing in his shoes and socks. Young laughed, until he realized how much worse it could have been. The man had suffered only bruises.

Dan Young settled easily into the factory, fitting in with the molders, machinists, and furnacemen, the blacksmith, stovefitters, and apprentices. He enrolled in an International Correspondence School course to study mechanical drawing and mathematics. Some of his drawings are still in the company files — and so, for that matter, are the patterns he made. They are stacked here and there, painted black or yellow, now mostly dust covered. Young also took a course in mechanical engineering, although he did not complete it. "I had all the books," he noted later, "and made good use of them when I had any engineering problems." The books are still at the foundry, too.

After he had been working at the pattern shop for some two years, the Ironworks was sold to the North Sydney company of A.E. Thompson but continued to be operated by the same workmen and managed by the same staff. Soon, however, things began to change. There were enough foundries and machine shops then in operation to tempt a man to try his luck elsewhere. This led to the constant traffic of workmen and engineers from one place to another always in search of better wages. Several men left Lunenburg for shops then offering the potential of greater income from piecework. Young's cousin, Arlington Rafuse, was one who left. He moved to Sackville, New Brunswick. In 1903, Young left too. He took a job in Amherst, Nova Scotia, at the Robb Engineering Company, where he worked in a 20-man patternmaking shop. Young learned about steam engines, for Robb built many of these, as well as sliding-valve stationary engines and various types of power-driven saws. It was a big, prosperous company that then employed 500 men. Soon after he joined Robb's, Young happened to step away from his bench to get a tool and a 10-ton jib crane broke loose from a supporting guy wire and landed where he had been standing.

Young's first assignment was to make a piston pattern. He asked what the record time was for such a job and did his best to surpass it. This was common, at that time, for the men took great pride in their work, and they could earn much more by turning out parts with both speed and precision. Speed remained a matter of pride for those who worked on an hourly basis, since the greater a man's skills, the more he earned. In the early years of the century, until World War I, in fact, $3.50 was considered a high daily rate, and only the most skilled could earn it.

Dan Young enjoyed his work at Robb's. It was only the fitting of grates into a boiler that he regarded as unpleasant, and for good reason. "One had to fit the grate patterns in the fire-box of a boiler with the riveters and caulkers working on the boiler. Even with one's ears plugged with cotton, it was quite an ordeal."

The workforce at Robb's was not particularly stable. Time and again, Dan Young saw men go off with their tool chests, bound for Montreal, western Canada, or the United States. Sometimes, a man would start a shop of his own. Young himself left Robb's in 1906. He moved to Quebec and found employment at Allis-Chalmers in Lachine. There he qualified for the $3.50-per-day wage by making a particularly intricate and difficult pattern. He spent a year in Lachine before returning to Robb's, which had lost most of its patterns in a fire, the great danger faced by all foundries.

Young had once helped to put out a fire at Robb's during his own employment there. The fire started early one winter morning, and before it was doused, Young's clothes were so coated with ice that he had to defrost himself in front of a stove so that he could undress. In November 1905, he learned that, except for the machine shop, Lunenburg Foundry also had been mostly burned, forcing the workmen to leave Nova Scotia for the United States. In 1906, when a new foundry was begun on the foundations of the old one in Lunenburg, Young finished his work at Robb's and returned home to become the chief patternmaker. He was then 23 years old, and he stayed in Lunenburg for the rest of his career.

Although he at once began making patterns for a new series of cast-iron stoves, Young also became involved in rebuilding the foundry buildings and equipment. The foundry building itself was erected on the foundation remaining after the fire. Young built the four-ton overhead crane installed there. He found a steam engine and boiler in an old tanning yard, bought it, and supervised its installation and the installation of other machinery in the pattern shop. As the foundry was being rebuilt, more men were hired, including Arlington Rafuse, who had returned from his job in Sackville to become foundry foreman. By 1907, Lunenburg Foundry was back in production, making stoves and some mill and ship machinery, using the few patterns that had not been destroyed in the fire. It was soon after production had begun that Young and his co-workers began to take increasing notice of the few gas engines they saw or those they read about. They decided to build one of their own.

"I knew very little about gasoline engines," Dan Young wrote later, "as at that time there was only one in use in a boat in Lunenburg. Mr. Rockwell [the company's manager] and I bought one, giving our personal note to do so. It was a two-cylinder Palmer engine."

The Palmer was taken apart and carefully studied, as were the catalogs of many other engine companies. Young did not plan to copy anyone else's engine, but he needed guidance for such things as compression ratio and port dimensions. The first engine was completed in 1908 — it was a two-port one-lunger rated, as a guess, at 5 hp. In February 1909, Essen Levy, a stovefitter at the foundry, bought this engine. He used it for 30 years — in his boat during the summer and driving a circular saw in the winter.

The prototype was followed by a 10 hp, and soon Dan Young was making drawings for a series of two-cycle engines. Since he did not have blueprint

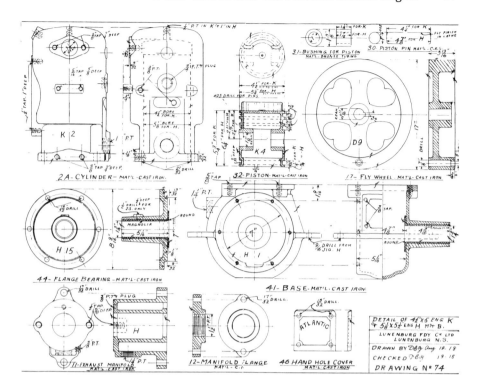

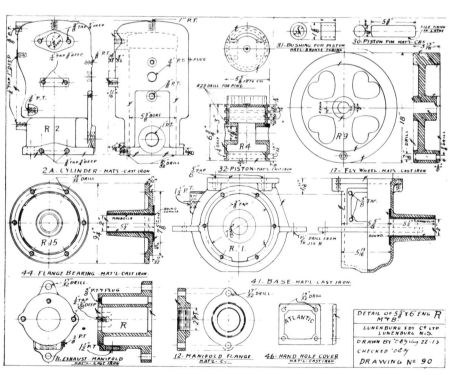

paper, he made the drawings on wrapping paper that he protected with a coat of shellac. These beautifully executed renderings were mostly drawn full size, although those for the largest engines were half size, making it easy to scale them up. The engines were a 3 hp (4 x 4), a 4 hp (4½ x 4½), a 5 hp (4½ x 5), a 6 hp (5⅛ x 5), a 7 hp (5⅛ x 5½), an 8 hp (5⅜ x 6), and the 10 hp with dimensions of 6 x 6½. Whether these were two- or three-port engines is now unknown, but a three-port one was designed in 1910. This was a 14 hp machine with three cylinders and bore/stroke dimensions of 5⅛ x 5½. None of these engines were engineered in the modern sense. They were the result of observation and trial and error. If there was a question about whether a part might be strong enough, the part was made heavier.

"I think our stuff was a lot heavier than anyone else's," Dan Young's son, Charlie, said. "If a casting was supposed to be an inch and a half thick, my father would say, 'Well, better make it 1⅝ just to be sure.' "

By the time the three-cylinder, 14 hp engine was built, Dan Young was well aware of the problems and potential of two-cycle design. One carburetor was used, and it was mounted under the exhaust manifold so that it would be warmed by it. To raise crankcase compression, he used balancers on the crank, filling up excess volume in the base. He also provided the base with one-inch air valves, which were interconnected, so that all three would open simultaneously.

"The air valves," wrote Young, "could be adjusted for each cylinder and . . . it was only necessary to open the carburetor needle valve a little more to compensate for the extra amount of air supplied the engine. The increase in rpm was considerable. We obtained the best speed running the engine 850 to 900 rpm."

This engine was installed in a 26-foot boat built by Obed Hamm of Mahone Bay. Experiments were conducted with differing port size, compression ratios, and propellers, and the results were measured on a course laid out from the railroad wharf in Lunenburg harbor to a buoy on Battery Shoal. Young timed his runs with a stopwatch. He measured the engine's rpm with a tachometer he had modified that was held against the crankshaft center hole. After a timed run, he would go alongside a schooner and, using the main throat halyard, raise the motorboat's stern out of the water, fit a different prop, and go out again. Eventually, the boat would do some 25 mph with an 18-inch-diameter, 36-inch-pitch, two-bladed propeller. At any speed, raingear was necessary, for the vessel was a wet craft. Young took her on a trial run to Chester and won a race against a motor cruiser whose builder had guaranteed a speed of 29 mph.

Subsequent Atlantic engines did very well in the informal races held by fishermen at their annual get-togethers. In 1935, company literature described how Angus MacNeill of Murray Harbor, Prince Edward Island, bested all contestants at the 1934 and 1935 Lobster Fishermen's Carnivals in Pictou. MacNeill was too busy with his 500 lobster traps to overhaul his 8 hp two-cylinder for the 1935 race or even to paint his skiff. He entered the race at the last moment and won. "Needless to say," said the brochure, "Mr. MacNeill thinks he has a good engine."

The company got its share of testimonials from those who didn't race their boats, too, since the production Atlantic engines were known as reliable machines able to withstand unusual abuse. Charlie Young recalls the experience of a fisherman caught off a lee shore without an anchor and unable to make headway against the wind and seas. He unbolted his 3 hp Atlantic make-and-break from its bed, took off the carburetor, and put the engine over the side on a long line. In this fashion, he rode out the storm. Afterward, he hoisted the engine back aboard. He drained the water, put his carburetor back on, bolted the engine in place, and started it. "That's unheard of," said Charlie Young, "but it's true."

During the early days of engine manufacture at Lunenburg Foundry, Dan Young was troubleshooter as well as chief designer. Sometimes he rode his bicycle on these missions, but often he had to take a train. He soon learned that dead batteries were a major source of trouble, so he usually took a battery tester and fresh batteries along with him. He was dismayed at how many trips he made only to find an empty gas tank or a throttle that had not been opened. When he wrote the instruction book that accompanied Atlantic engines, he advised owners who experienced starting difficulties, "Don't crank your head off." The manual then suggested that the operator check to see that the gas supply was working, the ignition switch was closed, the carburetor was adjusted, and the batteries were in good condition. Most of the genuine problems Young found involved maladjusted carburetors or ignition timing, or leaking head gaskets, for the first-series Atlantics were not domeheads.

Because few fishermen had ever seen a gas engine before they bought an Atlantic, most asked the factory to help install it. Dan Young installed engines in boats from Tancook Island to Shelburne. This is how he advised installing an engine:

> The bed pieces should be hardwood. To find the height of the bed pieces, stretch a line from the centre of the shaft hole to a point in front of where the engine is to be placed at the height of the shaft centre. This should allow the flywheel to be not less than four inches from the bottom of the boat.
>
> The first pieces should be placed cross-wise in the boat and must be well fitted to the planking and if possible resting on the keel. Bolt into or through keel and nail through planking from outside. There should be not less than two or three of these pieces. The pieces on which the engine rests must be lengthwise with the boat.

He suggested boring the shaft hole ¼ inch larger than the shaft and cautioned that the engine should not be bolted down until four strips of paper placed in the shaft couplings were held evenly. Often Dan Young got up before dawn when he was to install an engine and was still laboring in the evening. If something went wrong, he would miss his train home to Lunenburg. Once, in tiny Port Mouton, he was installing an engine in a small fishing boat. The owner shimmied up the mast to reeve a halyard. He had forgotten that the ballast was not in place and the boat capsized, submerging the engine and ruining the batteries. That night, Dan Young missed the train. He was never paid overtime for his work.

Nearly every component of the Atlantic was made by Lunenburg Foundry. When Charlie Young asked his father why the Atlantic had piston rings exactly ¹³⁄₃₂ inch wide — not a standard size — his father answered, "So our customers will have to buy new rings from us." Bemus timers were used initially on the jump-spark engines, but they were later replaced with a plunger-type crankshaft timer like Hawboldt's. The company bought Schebler carburetors for years, but after Schebler discontinued production of its long-lived Model D, Lunenburg made its own version. Crankshafts were purchased for years from a foundry in New Glasgow; cranks, however, were also made — and continue to be made — in house. These are machined from mild steel plates 2 to 2½ inches thick. The steel plate is cut to shape with an oxyacetylene torch, and the portion that will become the crankpin has holes drilled around its edges. The drilled portion is then knocked off with a sledgehammer, leaving the outline of the crank. Both ends of the crank are then turned to fit jigs for machining the crankpin, and the corners of the plate are rounded on an engine lathe, the steel removed in red-hot chips. Making a crankshaft like this requires some 5½ hours. It takes a good machinist 3½ hours to finish the crankpin and two hours to have the main journals prepared.

For many years the cylinders and bases of Atlantic engines were cast of gray iron. Eventually, Dan Young became impressed by accounts of the improvement that chromium and nickel could make in the iron, and he decided to add a small percentage of each. The metals were bought in shot form and, by experiment, Young and the foundrymen learned how to mix it. They used no pyrometers to check the pouring temperature.

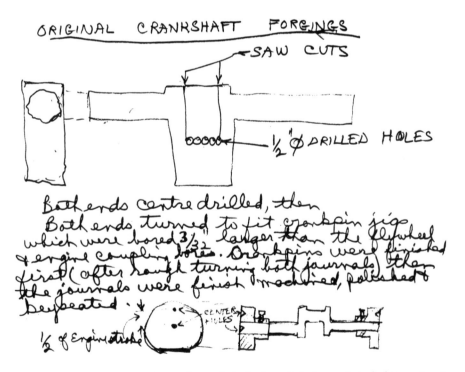

Sometimes Lunenburg Foundry made crankshafts from flat pieces of steel plate, cut out as shown and machined. (Drawing by Charles Young)

Steel plate billet and finished crankshaft, Lunenburg Foundry.

"They could tell by experience," said Charlie Young, "when the iron was hot enough. They'd wait until they could see sparks of the correct color flying out of it. They'd have the shot in a paper bag and pour that into the crucible and stir it around with a wrought iron rod. Then they'd be ready to pour the casting. But there was not too much science to it at all."

The company claimed that its engines were the first to have nickel and chromium added to the iron. In doing so, it noted that, although these materials cost 20 or 30 times more than iron, prices had not been raised. Since other engine builders also used nickel and chromium in their cylinder castings, the name of the first one to do so will forever remain unknown. Such techniques were not secret, and, despite warnings about not divulging company secrets, workmen moved so frequently from one company to another that nothing remained secret very long.

Just as the Atlantic engine was among the first to have improved cylinder materials, it was also one of the earliest two-cycles fitted with an aluminum piston. This happened, according to Charlie Young, because the fishermen of St. Pierre and Miquelon wanted to run their engines at rather higher rpm than normal. This was prompted by a desire to keep the upper hand in the fishermen's eternal duel with the Royal Canadian Mounted Police. The big dories were used, at times, to bootleg liquor into Canada.

These pistons were made of "saltwater" aluminum and were not liable to corrode. Clearance between piston and cylinder was doubled with the aluminum pistons, an increase of from .003 inch to .006 inch. "When they were cold," laughed Charlie Young, "they'd slap like crazy." Engines so equipped regularly turned 800 to 900 rpm instead of 500 to 600 and tended to vibrate considerably. Sometimes, the crankshafts were statically balanced in a lathe by the addition of iron weights. Improved aluminum alloys permitted the aluminum pistons to be standardized — with closer tolerances and without the attendant piston slap — shortly after the end of World War II. They weighed one-third as much as the cast-iron pistons they replaced.

By the time he was 16 years old in 1927, Charlie Young had finished the eleventh grade (the highest grade at Lunenburg's school), wanted to buy an automobile, and was ready to go to work in the foundry. When his father suggested that he attend college, Charlie Young said he would run away from home first. On July 4, 1927, his father took him into the machine shop presided over by his uncle Clarence.

"Forget his name is Young," Dan Young told Uncle Clarence about Charlie. "Put him to work."

Charlie Young began his career by sweeping floors in the machine shop for eight cents an hour, when the average wage was 20 cents. He worked each day from 7:00 a.m. until 5:30 p.m. On Saturday, he worked five hours. He realized that those wages were a slow way to buy a car and asked to work overtime on the few nights when he wasn't doing the correspondence course in engineering demanded by his father. After two weeks of overtime building stoves, he found he was not being paid for his extra work.

"Well," said his father, "you're going to be running this business someday, boy. You've got to work in every department. You've got to learn the business. You can't expect to get paid for learning the business."

In this fashion, Charlie Young learned to operate every piece of equipment at Lunenburg Foundry. It was not until January 1980 that a machine was installed that he neither knew nor cared to learn about — a computer. The Atlantic engines were built on a piecework basis, the workmen racing against each other and against themselves. Charlie Young always kept a small notebook in which he recorded all he did and how long it took him.

"July 26, 1929 — machined 8 6½-inch bells today. One hour 10 minutes each." He completed 20 flywheels for the 4 hp engine in 30 hours, machining the outside rim of the heavy casting on a lathe and then boring out the keyway. It took him six minutes to finish and polish a movable electrode. He made them by the hundreds.

This emphasis on how long it took to do things was important, since, among other things, it influenced the rather hazy manner in which the engines were priced. The selling price was based in part on manufacturing costs, but nobody really knew how much these were. Accurate costing was not a top priority. The most skilled of all engine fitters could turn out four engines in a good 9½-hour day, but Charlie Young recalls that prices were based on an output of two per day. Overhead was considered, the men's pay rate was mixed into the formula, and somehow a selling price was determined.

Despite his low pay and long hours of work, Charlie Young completed his correspondence course in engineering and also managed to purchase a 1921 Dodge Brothers touring car. He remained on good terms with his father, too. "He was a hard but fair taskmaster," he said, "but actually, he was more like a brother to me than a father." When he applied some of his newly acquired engineering skills to the foundry, Charlie Young received a shock. "I started figuring the horsepower requirements for the drive belts, and he looked at me and said, 'Why, I never did that.'

"I said, 'You mean you didn't figure so much per inch of belt width and speed and the number of plies?'

" 'No,' his father said, 'I never did that. I just decided a six-inch belt would be wide enough.' "

The largest two-cycle Atlantic engine ever built, and the last to be designed by Dan Young, was a 7 x 7½ three-cylinder rated at 54 hp. The foundry's sales manager, Dan Eisenhauer, had discussed such an engine with Dan Young, believing that its comparatively low cost — compared with that of a similarly rated four-cycle diesel — might create a market. Young felt the engine would be uneconomical, but when Eisenhauer sold a local fishing captain on the engine's virtue, the project was begun. Delivery was promised in six months.

"I started making drawings and ordered the crankshaft forging as soon as the drawing was completed," wrote Dan Young of this period in 1928. He and a patternmaker made up the patterns and drill jigs and ordered a two-inch carburetor from Schebler. "By good luck," wrote Young, "I struck the compression clearance correctly and did not have to make any changes, and even struck the propeller size, a three-blade 36 x 36 which the engine turned about 500 rpm."

Only five of the 54 hp engines were built. Charlie Young recalls that they were monstrous things and that he was afraid of them.

Other novel projects like the 54 hp two-cycle engine were sometimes begun at the foundry but not carried to completion. In 1920, oil-burning, semidiesel engines began to become popular in Nova Scotia, where as much as two-thirds of the Atlantic engine production was sold. Ignited by hot tube or hot bulb, these engines were cheaper to operate than a gasoline version, since the fuel was less costly. Dan Young spent three months making drawings for such an engine. He had completed the parts list and even made some of the patterns when the project was abandoned. The true diesel engine had already become increasingly practical for commercial marine use, and the expense of creating a new semidiesel no longer seemed worthwhile. Instead, Atlantics were offered, as were many engines, with a fuel/oil carburetor. It was fitted as an option in addition to the standard carburetor and proved particularly popular in Newfoundland. Two 20 hp Atlantics equipped with these devices powered a 50-ton schooner from La Have, Nova Scotia, to Hudson Bay, where the vessel was to be employed as a cargo carrier.

Engine sales remained brisk during the 1920s and into the 1930s. At one time, the Newfoundland business was accounting for 400 Atlantic engines per month, and the factory — which employed a maximum of about 200 men before World War II — added a night shift. The double-four model was the most popular. During those years, the design and manufacture of the engines remained unchanged. Among the few bought-out components were the coils, check valves for the water pump, petcock on the cylinder, and priming cup. The make-and-break parts, cam, timer, and trippers were made in-house. They were made of mild steel that was hardened by immersion in potassium cyanide. The resultant fumes were deadly and the workmen were careful to stay to windward during the process. Later they were made in "chill molds" in the foundry.

Continued in use throughout Atlantic production to this day is a hinged bottom end on the connecting rod. The arrangement has proved serviceable but has shocked more than one engineer. One day, an M.I.T. professor dropped

Cutaway Atlantic marine engine, The Fisherman's Museum, Lunenburg, Nova Scotia.
Inset: *Detail of cylinder.*

by the works and saw a cutaway engine on the showroom floor (the same engine pictured here). "Why, Mr. Young," he said to Charlie Young, "that won't work. You can't take up the bottom end of the rod. It's against all engineering principles."

"Well, that's a strange thing," Charlie Young answered, "because we've sold 12,000 of them and now we find they won't work!" In fact, there were shims permitting any clearance to be taken up.

The biggest single change to the engines laid out in 1908 occurred during the late 1930s. This was the change from the circular base to a split one. "We were having trouble getting the flywheels off the old ones to rebabbitt the main bearings," said Charlie Young. "See, there was no way to take them up at all, and when they wore, the engine had to be brought in for rebabbitting."

Getting the flywheels off proved a major undertaking. The wheel was usually rusted in place, the key stuck fast. Holes had to be drilled and tapped and special pullers devised for removal. Most flywheels had to be heated. Many flywheels were broken and many crankshafts bent in the process.

"Oh," said Charlie Young, "we had a lot of problems with them." He designed the new split base, keeping the volume as small as possible to ensure

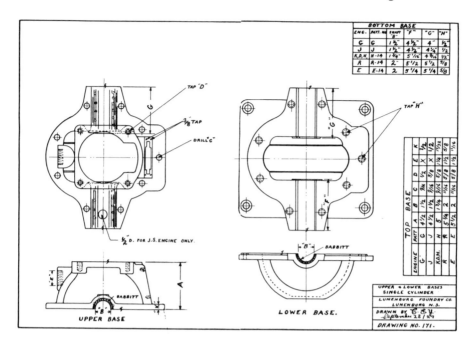

good compression. Balance weights were now fitted to the cranks to help dampen vibration and improve base compression still more.

During World War II, Atlantic engine production dwindled. The foundry became deeply involved in war work and employed 500 men outfitting minesweepers and destroyers during years when Lunenburg was home port to the Free Norwegian Navy. Most of the two-cycle engines built during those years were made by Acadia, according to the Canadian government's overall plans for who did what.

After the war ended, Atlantic production resumed, but on a much-reduced scale. The company brochure labeled the Atlantic "The Fisherman's Favorite," and it noted that the engine was "designed and built in a factory in Lunenburg where men KNOW the sea and what it requires of Marine Engines." A new-style Schebler carburetor was fitted after the war, and it permitted the engine to turn an extra 20 to 50 rpm at the top end, while providing a safe idle speed of under 100 rpm. The ideal of a really slow idle on the marine two-cycle was never really achieved.

As two-cycle demand dwindled, there was increased emphasis on converting automobile engines. These projects included the Ford Model A and various V-8s, a Chrysler six, and a Buda-based six-cylinder. As two-cycle orders dropped, prices for the Atlantics rose. They have increased tenfold since 1910, when a 4 hp could be bought for about $90. Although there has been talk about abandoning Atlantic production, this has not been done. A trickle of orders continues to arrive. In the summer of 1981, there was talk of an entirely new market opening up. Developing nations began to express interest in purchasing one-lungers. Possibly, there is still a demand, in the last part of the 20th century, for engines whose antecedents date back to the last part of the 19th century.

Dan Young retired in 1954 when he was 74 years old and died nine years later. After he retired, his son, Charlie, became the works manager, just as had been foreseen 27 years earlier when he first began sweeping out the machine shop. He reached 70 in 1981, a man like his father, able to do anything he needs to. He has designed two houses, including the one in which he and his wife live. Its plans were drawn in the hospital while Charlie Young was recovering from major surgery. In the basement is the fine darkroom where he processes film and makes professional-quality prints. Among the prints are those of the *Bounty* he made for Metro-Goldwyn-Mayer. The ship was built in Lunenburg for the studio, the engines installed by Lunenburg Foundry. Charlie Young was such a good photographer that the moviemaker offered him a job. He declined, but he kept the autographed portraits of the film's stars.

Charlie Young retired in 1976 as a vice-president and director, but when we met in the summer of 1981, he still remembered every board and tool in the foundry and the names of most of the staff. A bad knee, recently operated on, did not keep him from walking through the shops with me. He looked critically at everything that was going on, and it was clear that he could have stepped up to any bench and completed whatever task was being done there. We walked past the same machines that had bored out cylinders in his father's time and looked at the jigs on which the hinged connecting rods were mounted for installation of the bearings. Sometimes, Charlie Young would joke with the men and show off his scarred knee.

Connecting rod in place on jig used to install wrist-pin bushing. Note that bottom of rod is hinged, rather than bolted on both sides. Lunenburg Foundry.

"I've got a bone in my leg, the doctor says," he told them, and everybody laughed.

It was clear to see, however, walking with Charlie Young, that although the foundry looked the same as it did in earlier years, times had changed. He was unhappy with some of the men's long hair and unhappier still at the amount of dust that coated the hundreds of patterns in the pattern shop. He didn't say anything, but it was clear that he was thinking of how things used to be. He shook his head.

"Times have changed, Charlie," someone told him, guessing his thoughts. "Times have changed."

Charlie Young seemed sorry that they had. It was as if he understood that something important had been lost when the old gave way to the new, when the world of the one-lunger was overwhelmed by more modern times. He stopped for a moment by an ancient boring machine that was slowly honing out a 3 hp cylinder. Seeing the old machine, he almost smiled, but he did not seem sorry to walk out the door that afternoon and go back home.

The LOZIER engine has established an enviable reputation in yachting circles, and is now recognized as standard. We are too large a concern and have been in the manufacturing business too many years to make any careless or ill-advised statements which might prove misleading to the uninitiated, and we feel satisfied that the world-wide reputation and reliability of The Lozier Motor Company is a sufficient guarantee to all prospective purchasers that the statements in this catalogue may all be depended upon.

Lozier Gas Engines-Launches, circa 1909

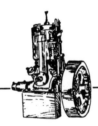

6

Daisy

BACK BAY IS SEPARATED from Lake Winnipesaukee by a low bridge. The bridge, on the main street in Wolfeboro, New Hampshire, is popular with fishermen. They lean over its stone walls to play their lures in the dark water, watching and listening for the boat traffic that plies back and forth beneath. When a boat approaches, they reel in their lines. Because the bridge is low, the only craft that can pass are motorboats. Back Bay is where Jim Doughty keeps his launch.

The first time I saw *Daisy*, she was drifting out of the sunlight. I stood in the slippery grass next to the water's edge and squinted, but I couldn't see anyone aboard. Then the morning silence was broken by a muted sort of putt-putt-putt, and Jim Doughty's head and shoulders popped into view as he took his place at the steering wheel mounted on the port side of the cockpit coaming. He was there for about a minute, heading toward the dock, when the motor stopped and he disappeared to crank it up again. It took rather a long time, that morning, to get *Daisy* from her mooring to shore.

Daisy is a 22-foot torpedo-stern launch built by the Lozier Motor Company in Plattsburgh, New York, in 1905. At that time, the company turned out such

Daisy *with owner Jim Doughty at helm.*

boats, and larger ones, by the hundreds. Each was equipped with a two-cycle or four-cycle engine of Lozier's own design and manufacture. The smaller boats got the two-cycles, the larger models — some were 50 feet or more — had four-strokes.

Once, there were many companies engaged in building and selling launches. They vied with each other for boat buyers' dollars, extolling the virtues of their particular craft — the wood, the design, the craftsmanship, the safety, the durability. Now all those companies are gone and so are most of the boats that they built. It is difficult to say how many launches survive or have been restored. *Daisy* lives on a mooring, like a waterborne dowager, surrounded by Chris-Craft runabouts. Lozier launches certainly are rare items. Jim Doughty knows of perhaps a half-dozen. Other Lozier buffs claim to know of more, tucked away here and there in barns and boathouses, but nobody believes the grand total could be much more than 12. Few launches have, as does *Daisy*, a Lozier one-lunger nestled between the engine bearers. *Daisy*, it is clear, is a rarity.

Jim Doughty has done his best to keep the old launch as authentic as possible. His unwillingness to alter her engine is one reason he experienced difficulty that morning, but occasional malfunction is something Doughty is willing to suffer. He insists on keeping *Daisy* as close as he can to looking and acting exactly like she did when she was built in Lozier's wooden boat shed, on the shores of Lake Champlain, over three-quarters of a century ago.

The Lozier Motor Company was organized by Henry Abram Lozier, Sr., in 1900. Lozier was born on a farm near Wrightsville in southern Indiana in 1842. At about the time the Civil War began, he opened a shoe business in Evansville but lost his capital to an unscrupulous partner. After that, he began selling sewing machines. He sold them so successfully that, in 1887, he invested some of his money in a bicycle-manufacturing company. By the time he was 42, he had reaped a vast fortune in the nascent bicycle industry, and he parlayed his investment into a four-factory empire. Company headquarters were in Toledo — the bicycles were sold under the name Cleveland — but the machines were marketed all over the world. Lozier was shrewd enough to foresee the end of the bicycle boom. In 1897, he sold out to the American Bicycle Company for $4 million. American soon began to lose money as sales dwindled.

Although he had formally sold the bicycle company, Lozier continued to maintain offices at the 11-acre Toledo factory where some 800 workers were then employed. It was at Toledo, under Lozier's direction, that his sales force and engineers continued investigating what Henry Lozier foresaw as the next great boom in transportation: automobiles and motorboats. While Lozier did his market research, George R. Burwell — the engineer who had designed the Cleveland bicycle — began investigating a number of European gas- and steam-powered cars. According to John Perrin, who joined Lozier as a draftsman upon completion of high school in Toledo in 1893, Burwell's work resulted in a motorized tricycle powered by a tiny De Dion Bouton single-cylinder, air-cooled engine.

The De Dion tricycles had created a sensation in Europe, where Albert De Dion, a wealthy count, had laid the groundwork for, among other things, mass production of motor vehicles and a series of road maps that later became the famous *Guide Michelin*. To Lozier, the idea of expanding from bicycles to tricycles seemed a natural-enough evolution. He built 100 trikes before he gave up his effort to popularize them in this country. Even as he did so, experiments with automobiles continued in Toledo. A number of steam-car prototypes were built, including at least one with a flash boiler inspired by the French Serpollet. All these automotive investigations ceased when Lozier voiced his continuing skepticism about the practicality of putting any of the machines into full production. Instead, he decided to build marine engines. There was a clamor for them, it turned out, and George Burwell had designed both a two-cycle one-lunger and a four-cycle twin.

Production of boats and motors was begun in Ohio, where both naphtha- and gasoline-powered launches were constructed. Henry Lozier's son Edward was frequently seen testing gas-powered boats on the Maumee River. Hulls were built in a plant in Bascom and shipped to Toledo on the Baltimore and Ohio Railroad. In Toledo, the engines were built and installed. Soon the naphtha portion of the business was turned over to the American Bicycle Company, and Henry Lozier made plans to expand in a big way. He sent another of his sons, Henry Lozier, Jr., to New York to investigate possible factory sites, for he planned to locate a factory in the great eastern marketplace and have access to ports from which Lozier boats could be shipped all over the world.

"The Lozier Motor Company of Toledo," said a press report, is "in correspondence with the civil authorities of several towns in the neighborhood of New York City with a view to establishing a large plant for the manufacture of gas engines, launches and automobiles."

A portion of the shop at Lozier. The traveling crane permits moving of heavy castings and equipment. The railway allows complete engines or parts to be moved from one building, or part of the shop, to another. At the left of the workman, one-lungers await installation of cylinder heads. Their pistons are visible. At the right of the workman, base castings await finishing and assembly. (Courtesy Morris F. Glenn)

Plattsburgh was the place chosen. Engineer Burwell had lived there once, superintending the manufacture of sewing machines. He thought the little city both pleasant and well located for building boats and marine engines. Henry Lozier was actively wooed by the town's bankers and businessmen, who subscribed 49 percent of the stock. Lozier owned the remainder himself. He was impressed with available manpower in Plattsburgh, as well as available water power. At nearby Indian Pass, he built a 1,500 hp electrical generating plant. According to Lozier, this power plant and its dam — built of timbers from Vermont and locally crushed stone — cost $50,000.

The electricity generated at the dam was used to power all the machinery in the new $350,000 shops. This equipment was described in the company catalog as being "of the newest and most modern type, much of which is especially designed and built for this class of work." The factories were equipped with traveling cranes driven by compressed air and served by Lozier's own industrial railway. The "class of work" referred to in the company's literature was high class — marine engines and boats aimed primarily at the pleasure market. Lozier never pretended that its one-lungers were meant for workboats. Instead, it viewed each one as an ornament to the Lozier launch it powered.

The changeover from the Toledo plant to the new site at Plattsburgh apparently was achieved by the spring of 1901. John Perrin remembers that he and George Burwell journeyed from Ohio to Plattsburgh — via Lake Erie and canals — in a 45-foot cruiser powered by a two-cycle engine. Perrin had rather quickly moved up through the company, progressing from draftsman to being in charge of much experimental work. Once the move to Plattsburgh was complete, he frequently tested boats on Lake Champlain and worked with Burwell to make improvements.

Lozier introduced its 1901 engines with faintly messianic prose describing the virtues of two-cycle engines and the evils of four-cycle types of small displacement. "The principal disadvantages of the four-cycle engines are in the fact that they must be built in larger sizes than the two-cycles developing the same horsepower, and every other stroke of the piston being an idle one necessitates extremely heavy flywheels and produces greater vibration. The four-cycle is necessarily much more complicated than the two-cycle, which alone is a great objection to it in a pleasure launch . . . . The absence of mechanical complications reduces to a minimum the possibility of derangements . . . ." For its largest engines, however, even Lozier chose a four-cycle. These were available in a 30 hp twin, 40 hp four, and perhaps in other models as well.

Lozier's two-cycle one-lungers were, like all of the company's products, of high quality. They were built entirely within the Lozier factories. Even the vaporizer, with its wheel adjustment device reminiscent of a steam engine, was designed and built in-house. Most Lozier parts carry the company logo. The crankshaft was made of drop-forged steel, the connecting rod of gray iron. The cylinders were carefully bored and ground so as to be interchangeable. Each was tested under 80 pounds hydraulic pressure. All bearings were bronze. Once assembled, the engines were run in off a factory power source before being started on their own and run, under full load, for six to 12 hours. They drove a propeller submerged in a specially constructed six-foot-deep steel tank.

"In this manner," said the company, "the engine is required to produce even greater power than when installed in a boat, and is thus tested in a practical manner with a view to performing the work for which it is built."

As if to enhance its credibility, the Lozier catalog was spiced with quotes from well-known engine experts like Gardner Hiscox and E.W. Roberts, both two-cycle fanciers. Quite on its own, however, the Lozier engine quickly achieved a good reputation. Gas-engine experts cited it as one of the most successful examples of the type, and in 1904, two Cornell students, G.H. Bayne, Jr., and E.C. Speiden, did a thesis examining the Lozier one-lunger for their degrees in mechanical engineering. At the time, the Lozier two-stroke was available in single- and twin-cylinder models of from 3 to 15 hp.

These engines were all two-port types, as refined examples as ever were made by any early marine engine company. As we've already seen in Chapters 3 and 4, the Loziers were fitted with a throttle valve mounted in the bypass passage, perhaps the best way to achieve speed control without interfering with the action of the vaporizer's poppet valve. The engines were also fitted with a muffler designed around a hot-air drum that conducted air to the vaporizer to warm it. A valve permitted the amount of this air to be regulated.

Rather than rely on gravity lubrication, the Lozier forced oil into the cylinder under slight pressure generated by the outflowing exhaust gases. The oil was fed through cast-in passages to the piston rings and combustion chamber. The rod bearing was splash lubricated, the oil poured into the base through a special fitting. The usual grease cups lubricated the main bearings. Ball-type thrust bearings, rather than discs, were fitted to increase the bearings' durability. The water pump was driven by chain off a sprocket on the crankshaft, a rather unusual arrangement on engines of this type and epoch. Ignition was make-and-break, the parts of which, like everything else on the engine, were made by Lozier.

The two Cornell engineering students tested one of these Loziers on a water-cooled brake wheel during five two-hour sessions. Every 10 minutes, figures were recorded showing the amount of fuel used, water used, rpm, temperature of the cooling water emerging from the water jacket, and exhaust temperatures. It was quickly determined that a minute adjustment of the vaporizer's needle valve significantly affected the proper running of the engine. A perfect mixture had to be attained for the Lozier to reach peak performance.

Because Bayne and Speiden tested the Lozier on its own and made few comparisons with other one-lungers of its power rating, their thesis stands primarily as an engineering document expressing figures of interest mainly to other engineers. They recorded the amount of heat generated per hour, the amount of oil used, etc. They noted that the engine did, in fact, develop its rated 5 hp. The comparisons that the students did make showed the Lozier to be a superior machine in some respects. It developed a mechanical efficiency — according to their calculations — of 79 percent, compared with the 69.5 percent efficiency of a Fay & Bowen two-stroke tested by other engineers and a Strelinger four-cycle, whose efficiency was found to be 76.6 percent.

The latter figure led the students to suggest that, although a four-cycle engine could develop power equal to that of a two-cycle of equivalent size, it would need to have much heavier parts. This, they pointed out, would increase materials costs. "We conclude," they wrote, "that for small launch service where the saving of weight and space is of some importance, the two-cycle machine will continue to hold the enviable position to which it has attained by its superior merit as a motor for pleasure boat service."

Just what Lozier had been saying.

The first Lozier launches were graceful craft with high, comfortable-looking coamings and fantail sterns. They ranged in size from 16 to 40 feet and were powered by engines of from 1½ to 8 hp. The 40-footer had a nine-foot beam and was capable of carrying 25 people. It cost $2,400, while the smaller models listed at from $400 to $700. These were the boats built in the Bascom, Ohio, shop. It was a large shop, but because the demand for boats was great, it always seemed overcrowded.

Like early automobile manufacturers, Lozier sometimes liked to photograph its products in front of a large sheet or dropcloth held aloft. The idea was to enhance the boat's — or car's — appearance by presenting it before a neutral backdrop. The method seldom worked as intended, however. The sheets always had wrinkles, and they were seldom long enough to form a backdrop for the entire subject. Thus, a 25-foot Lozier launch photographed with such a backdrop had both its bow and its stern sections protruding, and, instead of a plain sheet, one sees lengths of old pipes leaning against the wall and stacks of wood. Usually, Lozier dispensed with photographs for its early catalogs and used drawings instead.

Not long after the move to Plattsburgh, the boats were redesigned. The man responsible for this work, and for all subsequent design and construction supervision, was Frederick Milo Miller. Miller was a neatly mustachioed, bespectacled gentleman. He was born in Malone, New York, in 1870. As a young man, he went to Scotland to apprentice as a boatbuilder in Glasgow. When he returned to the United States, he worked for Charles Seabury from 1890 to 1893, when he opened his own shop in Malone. He joined Lozier when the Plattsburgh plant was erected. Four thousand boats were eventually built under his critical eye.

Perhaps the most important single innovation Miller introduced to Lozier was a change from fantail to torpedo stern. The shape of these sterns varied and included what most would now think of as a canoe stern.

"We were among the very first designers of this improved type," said the company of its torpedo-stern boats, "and the fact that the most noted and able designers of power boats have since pronounced in favor of the torpedo stern is a justification of our faith in the superior qualities of this type of launch."

The 1901 catalog included comparisons of both fantail and torpedo-stern boats viewed from the bottom. The illustrations revealed that the fantail model had a substantially shorter waterline than its successor and that the torpedo stern's beam was greater throughout the after sections. This permitted the engine to be mounted well aft and thus left more cockpit space for passengers without causing the stern to settle.

"It is absolutely impossible," said Lozier, "for this model of boat to settle or drag, even at full speed . . . . The overhang in the fantail type of boat is practically valueless in giving the boat support." The company also claimed a 1 mph speed advantage for the torpedo-stern boat, a figure arrived at after testing boats of equal length with identical motors.

These Loziers were strongly built of fine materials. Keels, stem and stern pieces, keelson and deadwood were all made of air-dried white oak cut in Lozier's own sawmills. Forged bronze bolts secured the backbone together. Frames were oak, too, closely spaced at from four to seven inches, depending on the size of the launch. The frames were beveled to give firm support to the cypress planks. Cedar was used if specified by the customer. Planking was fastened with copper rivets set into counterbored holes and covered with plugs. Decks and coamings were oak. Floors were clear white pine, covered, if desired, with linoleum. The copper fuel tank was contained in the bow in a watertight compartment. Safe storage of gasoline was a topic of much discussion and concern, so each tank was pressure-tested for 12 hours. The fuel line was routed outside the boat, along the garboard, where, owners were assured, it was quite protected from damage if the boat ran aground.

Each of Lozier's models was given a name. There was the 17-foot River Special, the 21-foot Outing Special, the 25-foot Lake Special, the 31-foot Club Special, and the 31-foot Hunting Cabin Cruiser, the latter intended for family cruising. The 36-footer was the Open Water Cruiser; the 41 and 50 shared the title of Atlantic Cruiser. Largest was the 62-foot Revenue Cutter, which could be powered with two 20 hp engines or a single-screw 40 hp unit. A dummy smokestack was fitted. It actually held hot water for the bath(!), but nonetheless gave the Revenue Cutter the appearance of a fancy steam yacht.

Lozier's boat and engine business prospered between 1900 and 1905. A second factory was opened on the East River in New York City. There were branch offices in Europe and the Far East. A Lozier executive, Fred Chandler, set up the business in Germany. Later he would build automobiles under his own name. The boats were raced actively and successfully in many parts of the world. In 1904, H.A. Lozier, Jr., piloted his *Shooting Star* to victory in the prestigious Gold Challenge Cup.

Automobiles had not been forgotten at Lozier. Several engineers within the company who had worked on the early experimental cars clamored to begin production. Among them were Perrin, Burwell, and another engineer, J.M. Whitbeck. Before Henry Lozier, Sr., died in 1903 — of a heart attack in his rooms at the Waldorf-Astoria in New York — he had consented to further research into possible automotive production. Perrin went to Europe to look at what carmakers were doing. To disguise Lozier's plans, Perrin resigned from the company. He filled a series of notebooks with details about what he observed in Europe's finest automobiles, including Panhard, Mors, Mercedes, and others. Then he returned to America, rejoined Lozier, and engineered a four-cylinder automobile engine and a 115-inch-wheelbase chassis. The first production Lozier car was exhibited at Madison Square Garden in New York in 1905. A paved one-half-mile stretch of road was built at Plattsburgh for testing. Corral Lewis, whose father worked at Lozier's as a toolroom foreman and earned $19.75 for a six-day, 10-hour-per-day week, recalls that it was the first paved stretch in the area.

Once Lozier entered the automobile business, rumors began to circulate that the company would discontinue building boats and marine engines. This story seems to have begun in the summer of 1906, which is when *The Rudder* mentioned it, and it eventually proved to be true. By 1907, Lozier was building 50 or 60 cars a year, while boat production had greatly diminished. Ralph Mulford, who had joined the company as a young man of 16 in 1901 to repair and test the launches, now became a racing driver instead. He piloted Loziers in most of the great races of that era and placed third at the first Indianapolis 500, held in 1911.

In 1908, Lozier's boatbuilding activities were transferred to Traverse City, Michigan, to make more space for automobile production in New York. When the Plattsburgh factory had reached a strained capacity of 600 motorcars per year, a new factory was opened in Detroit. Harry Lozier asked architect Albert Kahn to design the new facility. It was a 65-acre project that was ready in 1911.

Lozier finally ceased its marine operations around 1910. Nobody seems certain of the date anymore. Because boats, not cars, were Frederick Milo Miller's life, he established his own boatshop in Plattsburgh and called it the Lake Champlain Boatbuilding Company. It was not a great success, and he gave it up in 1912 to become a teacher. The Plattsburgh factory, where he had once supervised boatbuilding, continued producing automobiles, but, by 1915, it had fallen upon such hard times that both John Perrin and Harry Lozier resigned. The factories were sold in March 1915, and the last Lozier car was built three years later under a reorganized, refinanced company that failed. Both Harry Lozier and his brother, Edward, worked in the automobile industry after they left the company that their father had begun. Neither one regained the fortune that had been lost. For a time, Ed Lozier worked at Jeffery — later to become Nash — at a salary that was said to have been about half of what he ordinarily tipped a waiter when he had been a wealthy man.

The narrow road into Jim Doughty's weekend place winds among pine trees. There is one particularly tight turn, and when you round it, you see a red, white, and blue sign nailed to a tree — Interstate 95. Doughty's cottage stands a couple of hundred yards farther down the dirt road. It is surrounded by trees. Under several of them, old motorboats lie in cradles, covered with tarpaulins. There are two workshops. In one of them rests a Lozier launch that predates *Daisy*. It is a fantail model to which somebody has added a cabin. Doughty plans to restore the boat, just as he did *Daisy*.

Jim Doughty.

Before *Daisy*, Doughty owned a 1952 Chris-Craft runabout. He restored the boat himself and kept it in the Wolfeboro boatyard run by George Johnson. The place is a kind of Chris-Craft heaven and a clearinghouse for many of the boats as they come on the market. It was Johnson who put Jim Doughty in touch with *Daisy's* owner when the old Lozier was about to be sold by the same family who had owned it for many years, on Lake Champlain.

"We took a weekend trip to see the boat," Doughty remembered. "We had high hopes and a couple of Polaroid pictures the owner had sent us. We bought the boat."

The Lozier's graceful hull had been sheathed in fiberglass. The original engine had been replaced by a Model A Ford unit. Doughty took out the engine and began removing the fiberglass. "It came off okay," he said, "but the resin remained. What chemicals couldn't get off, I had to sand and scrape." He replaced part of the keel and half of the ribs. The rotted portion of keel surprised him, because, when he had inspected the boat, he had been unable to poke his knifepoint into the hardwood. When he got home, he found he had been prodding cement. After he removed the cement, he found the rot. The decks were entirely replaced. The cypress hull planks, however, were in good condition. He routed the seams and recaulked. "I worked on her for three years of long nights and long weekends," Doughty says.

Although he sold his Chris-Craft, Jim Doughty remained in close touch with George Johnson, who later learned of an available 3 hp Lozier engine. It had belonged to an engine collector in upstate New York. When he died, his engine collection was sold. Doughty bought the Lozier, sight unseen, for $500. It did not even need a paint job.

"I never dreamed," said Doughty, "that I'd ever find a Lozier. The only two I knew of were up on the St. Lawrence and they were in the museum at Clayton. Later I found out about another one in the Finger Lakes region and about some parts. If you need a part, all you can usually do is pray or have one made." He has only a few spares, among them a coil.

Although he now owned both a Lozier launch and a Lozier engine, Jim Doughty found he was short of information on precisely how the two were to be joined. He culled methodically through all the Lozier literature he could obtain, noting details of the engine and the suggestions from the company for installing it. Because the pages of one catalog were stuck together, he did not know about the hot-air drum that sends exhaust-heated air from the muffler to the vaporizer. Later, when the engine proved balky and unwilling to run while cold, he began asking questions. But there was nobody he knew of who could be of any real help. He went through his catalogs again and again.

Because of the problems he associated with the engine's vaporizer, Doughty considered replacing it with a Schebler carburetor. He refrained from doing so out of determination to keep the engine original. Besides, the vaporizer mounted on threads that were unlike the standard threads required by the carburetor. It was as if Lozier had devised its own threads just to make things more difficult. Finally, one day Doughty unstuck the pages of his catalog and learned how the vaporizer was supposed to be warmed by hot air from the

Lozier one-lunger.

muffler. By then, the launch had already been fitted with an underwater exhaust, not a muffler, so Doughty "freelanced" a preheater arrangement. He located a prop shaft with a reversing propeller in Brookfield, New Hampshire. The shaft fit the boat perfectly, according to Doughty, but looking beneath the floorboards at the machinery, one gets the impression that installation must have been more complicated than that.

Daisy does not have one of the original Lozier fuel tanks, although Doughty says he knows of one and thinks he may try to buy it. The company offered several tanks. The standard version was a cylindrical model, but those who wished to have additional capacity could order a tank shaped specifically to the bow section of their launch. On *Daisy*, the fuel line no longer runs outside the hull. Her owner is less worried about the results of a fuel leak than he is about additional holes in the boat's bottom.

As his restoration project neared completion, the only element Doughty despaired of finding was one of the handsome bronze builder's plaques that

Reversing prop shaft in Daisy.

Lozier had mounted on each of its boats. The only plate he knew of belonged to a family of Lozier enthusiasts in Plattsburgh. They loaned him the plate, and a jeweler and old-boat hobbyist named Syd Marston made an exact replica for *Daisy*.

"He wouldn't take a nickel for his work," said Doughty. "That's how things often work in the hobby. People can't wait to help you out."

Gradually, after much trial and error, *Daisy* began to near completion. The original steering cable, made of tinned copper or brass, had stiffened up and was replaced. The brass wheel and the steering winch drum, the latter turned from mahogany, are original. Finally, the old Lozier was complete. She was launched into Lake Winnipesaukee and the sunlight sparkled on her white hull

Special builder's plate cast for Daisy.

and golden varnished oak topsides. She reminded Doughty's wife, Priscilla, of a daisy, and that was how she was named. The boat has been awarded several trophies at the annual antique boat meets held each summer on the lake.

Living with a technology that will soon be a century out of date has its trying moments. Jim Doughty has had plenty of chances to ponder this, but he tells himself that having things go wrong is, after all, merely a part of the old-boat game. "The first year I had the engine," he remembered, "I had all sorts of problems. It is very, very temperamental." The old one-lunger soon developed a personality whose major trait was moodiness. "When it gets in one of its moods," said Doughty, "I don't care what you do. It just won't run. That vaporizer," he said as he held his hands palm upward and searched for the words, "is something else."

By the time Jim Doughty became the engine's owner, a lot of hands had passed over its gleaming exterior and had delved into its mysteries — to little advantage. Doughty found the vaporizer to be a source of innumerable maladies, each stranger than the one before. He took apart the device and found that, instead of just one spring inside to seat the valve, there were three. Which one was right? Which had the correct tension? He noticed that the valve guide was worn, too. He lapped the valve in its seat with compound and put a new bushing in the guide. He chose the one spring that appeared most "right" and assembled the vaporizer. He did all this and built the preheater, too. It all helped, but it did not guarantee a reliable machine.

No matter what has been done to the engine thus far, it usually seems to require attention of one sort or another. Sometimes, its moodiness finally gets to Jim Doughty and he walks away from the boat. In the summer of 1980, when it was over 100 degrees out on Lake Winnipesaukee, the engine stopped and he could not start it again. Starting it requires muscle and optimism at any time, and that day all of Doughty's optimism drained away. He paddled *Daisy* home in disgust. Later, when he returned to examine the engine, he found the trouble to be two loose screws in the timer. "I was out there all that time," he said, "and I never noticed those two screws." Because the igniter hammers away incessantly, every nut and bolt in its linkage must be perfectly tight. The result of a loose nut is an inoperable engine.

The day Doughty and I went out in *Daisy*, the engine was feeling moody. The boat looked beautiful, her brass polished, her varnish gleaming, but the engine seemed to resent having someone aboard who had come to poke and prod and photograph it. It was a warm morning in July but not warm enough, apparently, for the vaporizer's liking. By the time *Daisy* finally reached the dock, Doughty was shaking his head.

"I think this is one of its moody days," he said. His daughter, Melissa, had gathered a pile of equipment on the dock, just as if things were going to work out just fine.

"This used to be a clean T-shirt," said Jim Doughty, looking down at the grease stains. He knelt before the engine, reached for the flywheel starting handle, and heaved it upward. The Lozier came to life immediately. With its

Starting Daisy.

underwater exhaust, it ran relatively quietly, with a sound like many pieces of steel whispering to each other. For a few moments, *Daisy* strained hopefully at her docklines. Then the engine slowed and abruptly died.

"Maybe it just needs to get warmed up a bit more," I suggested, hoping the words would make us all feel better.

"Maybe that's what it is," said Doughty. He took the gear as I passed it down from the dock, stacking it on the tan seat cushions. During the next 15 or 20 minutes, the engine ran many times, but never for very long. Doughty removed the stationary electrode, which, on the Lozier, is mounted in the center of the cylinder head. He put a screwdriver down the hole until it rested atop the piston. Then he rotated the flywheel, watching to see when the igniter broke contact. The engine was timed perfectly. Almost regretfully, he screwed the electrode back and wiped his hands with a rag.

On the assumption that the engine would cooperate more if we cast off from the dock, thus displaying our faith, we started the machine again. The Lozier thumped powerfully and we breathed sighs of relief and settled back to enjoy the ride. The way one of these old launches moves through the water has to be experienced to be appreciated. *Daisy* sliced silently along, leaving little wake and riding smoothly. "With these old displacement hulls," said Doughty, "you don't get your brains beat out." We had just begun to really savor the boat when the engine stopped again. "Miserable," Doughty said.

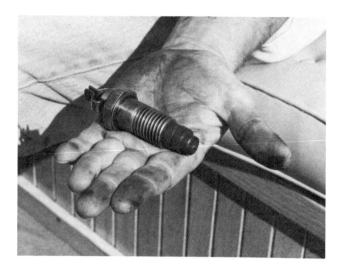

Stationary electrode of Lozier engine.

That sort of thing, unfortunately, characterized the whole morning, but we did not give up easily. Once, I happened to be looking right at the vaporizer when the engine died. "Did you see that?" I asked.

Jim Doughty shook his head. "That time, when she stopped, a puff of blue smoke came right out of the vaporizer." It was as if the valve were not seating properly. We went back to the dock and took off the vaporizer's top. Things looked normal inside. Doughty performed a few minor adjustments and we tried again. That day, the engine never did run for longer than 10 minutes at a time. Doughty was most apologetic and obviously unhappy. I got the feeling that he finally was tired of the Lozier's moodiness and that he contemplated something drastic — a newfangled Schleber carburetor, perhaps.

"Don't do anything drastic," I told him. I had been stranded too many times by old boats and old cars to be much surprised by the Lozier's performance.

"I'm going to do something," Doughty said.

It was a couple of weeks before I talked with Jim Doughty again. "How's the boat?" I asked.

"Fine," he said. "After you were here, I took that vaporizer off and took it down to a machine shop. We found out the spring was not seating the valve reliably all the time and made up a bushing to insert in the valve guide." He sounded relieved, as though he and the old Lozier had at last come to terms. The last time we spoke, *Daisy* had just been hauled for the winter. All through the autumn, her 75-year-old engine had run just fine.

He is completely infatuated with gasoline motors and collects them from the most improbable places. Once he brought home an old motor that someone had hauled into South Arm for a mooring anchor and then abandoned. It had been under water all summer and frozen into the ice all winter, but he dragged it the seven miles home on a hand sled, brooded over it, took it to pieces and put it together again, and now it runs the saw that saws our firewood.

Louise Dickinson Rich, *We Took to the Woods*

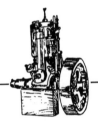

7

The Collectors

THERE IS NO MISTAKING the house where Chris Hawes lives. It is a large wooden structure on the same road that leads down toward lush meadows and the Minas Basin. The meadows here were reclaimed from the sea by the Acadians in the days before they were uprooted by the British and forced to leave Nova Scotia. Chris Hawes's place is not far from the little church that commemorates this sad exodus. His is the house with all the old engines in the yard.

Should you ever go to Grand Pré, you might see him out there, starting up a big stationary engine for the delectation of another enthusiast or to impress some nonbeliever. "There are a lot of things," Chris Hawes will tell you, "that don't have half the fascination of a couple of tons of steel revolving at 60 rpm with everything synchronized and running properly and performing a real valuable function as far as the development of the entire planet earth is concerned." Chris Hawes is a man who doesn't take gas engines for granted.

It is difficult to estimate the number of old-gas-engine enthusiasts. *The Gas Engine Magazine,* founded in 1966, has subscribers all over the world, and, each month, about 16,000 copies are mailed out from its Lancaster, Pennsylvania, offices. "An internal combustion historical magazine," it calls itself.

The want ads include everything from old farm equipment to marine engines, hot-air engines, and road-building machines. In one issue in 1981, a collector in Luxembourg placed an ad looking for Lenoir engines.

Many of the magazine's readers have formed clubs, and a loose network has been created for the swapping and selling of engines and parts and the exchange of technical and historical information. There are, however, many enthusiasts who do not subscribe to *The Gas Engine Magazine*, people who are somehow involved with engines entirely on their own. There is simply no counting all of those who may have one or more of the old machines out in the garage or shed. When he learned of this book, a friend of several years suddenly informed me that he had an entirely original, operable stationary engine in his garage basement. We clambered down the stairs, and there, beneath a crumbling old wooden canoe, was an Alamo farm engine built sometime in the 1920s. "I've never known quite what to do with it," my friend said, "but I've never wanted to get rid of it, either."

Chris Hawes bought his first engine one day in 1969 when he went out to purchase a stove. Right there in the stove showroom was an old Acadia marine engine, sitting on the floor, and Hawes experienced something that must have been akin to love at first sight. He bought the Acadia for $5.

"Since then," he says, "it's all gotten a little out of hand." His wife nodded in agreement.

Not long after he acquired his Acadia, Hawes made a pilgrimage to Bridgewater, Nova Scotia. In those days, the company still had time to show people around the old works. Chris Hawes climbed the wooden steps to the attic, steps worn concave by generations of Acadia employees, and was allowed to comb through all the material deposited up there. It was a clean place, as attics go. It was full of faded photographs, old catalogs, and cartons overflowing with the wooden blocks once used to print them. There were even some of the embossed metal signs that used to denote one of the company's dealers. "*We Sell Acadia*," they announce in glossy gold and red letters — "*Always Dependable*." A snub-nosed golden speedboat plows across a deep green maple leaf. Attics like this are prime territory if you like gas engines, and over the years, Chris Hawes has done his best to get into every likely looking attic in the Maritimes.

The results of his searches are evident throughout his big yellow house. There are shelves full of rare old books about gas engines, drawers teeming with antique photographs and sales literature, and there are the engines themselves. They are in the yard, in the cellar, in the living room itself. They repose in places where one might otherwise expect to find an end table or a big vase full of flowers. They sit on varnished wooden bases like brightly painted bits of sculpture.

Although he had once tinkered with the Weber carburetors of an Austin-Healey Sprite, Chris Hawes does not consider himself to be mechanically inclined. As he studied his Acadia, however, he concluded that it was simple enough for even a layman to understand. He found, in fact, that there was

Chris Hawes's living room. Marine engine is a Hartford.

nothing to keep the little engine from running again. As he cleaned and painted it, Hawes developed that same deep affection for the machine and what it represented that characterizes all those who cherish old engines. And he found that he enjoyed putting old engines right again. That was when he began buying one old engine after the other, until they overflowed from his house into the yard. To him, the engines present "a very strong visual association, something you wouldn't get with a car or motorcycle engine. You couldn't very well collect a six-cylinder Chevrolet engine, for instance. These old gas engines are entities in themselves."

Because a marine engine is somewhat less of an entity unto itself than a stationary engine, it is less popular with collectors. It is not as easily started and operated as a farm engine, and, because much of the old-engine movement has always been an inland sort of thing, marine engines are best known to collectors in coastal areas and around the Great Lakes. One old farmer, seeing a marine engine at a show, puzzled over it for a time. Then he asked the owner, "How come this engine is catawampus?"

"Huh?" said the owner.

"What for is it tilted down like that?"

"Because it's for a boat."

"Oh," said the farmer, "a boat." He walked away, shaking his head and pondering the marvel he had seen. "A boat."

Chris Hawes at site of his planned museum, June 1981.

The serious collecting of marine engines is one of the more recent phenomena of the old-engine world and of the world of antiques as a whole. "There are not so many marine engine collectors around," Chris Hawes believes, "because the engines are single purpose, and they are a bind to set up and run. I don't think they'll ever attain the popularity of stationary engines, but they do often look awfully nice."

Hawes has collected all kinds of old gas engines. He says that he does not like to think anymore of all the hours, days, gallons of gas, torn clothes, and even the occasional enemy made during his years of aggressive searching through attics and barns. But the collecting bug has bitten him deeply. He traces every lead he hears about, searches diligently through the storerooms of any company that still survives and once built engines, and regularly answers phone calls and letters from collectors elsewhere. He is very much in competition with his fellow enthusiasts, proud of his information sources, and always looking for new leads.

And yet, despite his strong acquisitive urge — one that Hawes shares with other serious collectors — his feelings for the machines themselves run deep. His enthusiasm is boundless, not only for the engines but also for the companies that built them and the men who once used them. He understands that the engines represent a tangible link with a past that, in many ways, seems more appealing than the present. He knows, too, that unless the engines are saved, unless the stories of those who built them are set down soon, the link will be broken and something precious will be irrevocably lost.

These are the reasons that Hawes plans to create a museum commemorating the engines and the world they represent. He has moved an old wood-peg barn from Ontario to Grand Pré to house an engine display. Engines will be set to work. They will run saws to cut shingles and presses to make cider.

"I'll be curator," said Hawes. "I want everything to run and I want to bring other old buildings here and make something like the Ford Museum's Greenfield Village."

It is individuals, not museums, who have thus far done the most to preserve old gas engines. One need only go to one of the collectors' gatherings to see how exciting it can be to stand in a field full of brightly polished old engines, each thumping away just as steadily as it did 50 or 75 years ago. By contrast, few museums have done anything more with the old engines they acquire than to leave them untended in a storage area. If they are marine engines, they are likely to suffer damage as their iron cylinders continue to dry out and crack.

A severely rusted water jacket eventually turns to ocher-colored powder. When scraped onto the floor, this detritus looks like the leavings of hungry termites. Once it appears, it often means that the old machine is no longer salvageable. Dried-out water jackets are one reason why some collectors have such fascinating workshops. It takes a lot of machinery and skill to restore a one-lunger to life. Such engines often need much more than a wire brushing and a new paint job. They need a craftsman as able as those who built the machines in the first place.

Richard Day has been collecting old marine engines since 1935, when somebody told him about a Lackawanna that was for sale. All he had to do with that engine was to clean and paint it, which did not require the facilities of a very fancy shop. But that was a long time ago. Now the basement of Day's Maryland home is bursting with old marine engines. They stand about virtually everywhere, tucked into corners and under shelves. They surround the Model T roadster in which Day and his wife, Barbara, used to go for drives. They sit in neat rows under dust covers or perch singly on shelves. In the little

Dick Day in his Severna Park, Maryland, basement with crankcase and crankshaft of two-cylinder Palmer.

space left over are some of Day's old radios. On the morning I met Day, he was busy unloading the pieces of what was then his latest old engine from the back of a pickup truck. It was a two-cylinder ZR series Palmer. He hefted the bright green flywheel. "I don't know where I'm going to put this," he said.

His basement was so crowded that he had to move boxes of parts just to get his workbench cleared away. The bench was not far from the prize lathe that he salvaged from a Baltimore boatyard. The lathe was built sometime in the 1880s by the F.E. Lee Company of Worcester, Massachusetts, and it was covered with filth and about to be sold for scrap when Day purchased it, cleaned it up, and installed it in his basement. It seemed to be a most appropriate sort of lathe with which to restore old boat engines. Over the years, Day has acquired a number of other antique machine tools. He has also taken a course in machine-shop practice to learn the fine points of operating them.

Besides his many tools, Day has gathered a myriad of bits and pieces, bolts, screws, pipes, and cotterpins. "As you can see," he said, waving idly at the basement, which is stacked from floor to ceiling with shelves and treasure-filled cigar boxes, jars, and cartons, "I've got all kinds of stuff. Most of this stuff I got at auctions."

He pointed to some large drill bits. "Who wants stuff that big? Nobody. But a drill like that new would be worth $45 minimum. I got one the other day for 50 cents."

One cigar box was full of high-grade cutters of all sizes, made of high-speed steels. He had a fortune in cutters in the cigar box, but they cost him only three dollars. He moved aside a carton and picked up a box full of taps and dies made in Israel and in Spain, tools of a quality rarely found at all, let alone at a reasonable price. Not far away were boxes containing nonplated cotterpins. To a layman, they would look like just plain, everyday cotterpins, but Day recognized them as a special treasure.

"For restoration purposes," he said, "they are invaluable. You try and buy a nonplated cotterpin and you have to order them by the barrel. When the old engines were made, however, unplated cotterpins were what was being used. They are the only ones truly authentic." He is a stickler for accuracy and would no sooner put the wrong sort of cotterpin on an old engine than he would an inappropriate carburetor or a pipe with the wrong sort of threads. To do so would be an insult to the whole idea of the hobby, an affront to the memory of all those who once designed, built, or worked the machines. Like Chris Hawes and those others deeply committed to old gas engines, Day would not think of thus desecrating them. He might not necessarily come right out and say this, at least not until he knew whether or not he was talking to someone who would understand, but he and others feel this way. Theirs is a deep and abiding affection for the old iron, and they are careful to avoid acts of sacrilege.

Among the most sacrilegious acts of omission one can perform regarding an old marine engine is to let it "dry out." Drying out occurs when the iron castings become porous and soft. Salt residues left in the water jacket pick up moisture, and the iron slowly sloughs off. Very gradually, the engine turns to mush. Cracks develop, and, from within them, the dread ocher-colored iron

Two-stroke, circa 1914, with timing advance device clearly visible. Owned and restored by Dick Day.

can be chipped off with a thumbnail. Sacrilegious or not, old marine engines are permitted to dry out all the time. Usually their owners simply are not aware of what is happening or else do not know how to prevent it. An old engine needs to be filled with oil or antifreeze if it is to be preserved. Once the iron begins to decay, repairs are difficult, for welding the iron is a touchy business. Usually, afflicted areas must be cut back ruthlessly and new pieces welded in. After such surgery, few engines ever run again.

Although each engine presents a different challenge and some are more easily restored than others, Dick Day has evolved a general scheme of restoration. He uses a wire brush to take the engine down to bare metal. If necessary, he will sandblast it. He thoroughly cleans the pieces in gasoline, although he worries about the possibility of fire while he does so. Then he repaints the castings with epoxy-based paint. This yields a tough finish that is easy to repair if a chip should occur. He uses epoxy on the inside of the crankcase, too, rather than the red lead used by the old-time manufacturers.

"I use it," he said of epoxy, "because it bonds to the metal so well."

Because the paint is expensive, he often uses rather unusual colors for the base coats, colors available at bargain prices. Day uses a wire brush to clean all the nuts and bolts he removes. If the bolt heads will not be painted, he protects them with either clear plastic floor wax or clear lacquer. Because rust is

1914 Lackawanna four-cycle owned and restored by Dick Day.

an ever-present threat in his basement, he keeps two dehumidifiers running during the steamy Maryland summers.

One metal Day never wire-brushes is brass. He has a special affection for brass, a connoisseur's eye. He may hold a brass object up to the light, gazing at it as if it were a glass of wine to be checked for color. "A lot of modern brass," he said, "is almost white in color instead of yellow. Old brass has a different shade. Turn-of-the-century brass is different." He cleans brass parts with a rotary disc and rouge. "See this?" he asked, picking up a tarnished old mixing valve from its place on a shelf. "When this is all polished, the brass will look beautiful."

Two-cycle mixing valves, before and after. Owned by Dick Day.

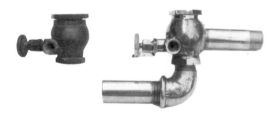

Having been a collector of old marine engines for nearly four decades, Day has watched the hobby grow slowly until the recent modest explosion of interest. With it has come an explosion in prices that has already begun to narrow the field. These days, it has become an unspoken law among many serious hobbyists not to talk money but to ask instead what the other man has to swap. It is not often, anymore, that Dick Day repeats an experience he once had with a longtime Palmer dealer on Maryland's Eastern Shore.

"One day," he said, "I went over to Kent Narrows to see Captain Eddie Sevier. I asked him if he had a manifold for a particular kind of Palmer. He didn't answer me for about 15 minutes. He was thinking about it. Then he said he didn't have any used manifolds but that he did have some new ones, although they were expensive."

For a moment, Day wondered if the explosion in old engine prices had touched even Captain Sevier's broken-down old chandlery at Kent Narrows. He asked the old man how much the manifold would cost, and Captain Sevier let about 10 more minutes slip past while he went about other jobs.

"It'll cost you," he piped up at last, "six bucks."

Dick Day had been prepared to pay $50 and was relieved. He was more relieved when Captain Sevier added a new gasket in the deal. "You'll be needing that," he told him. He also sold him two flywheels for 50 cents apiece. "The whole thing," said Day, "was unreal. By the time I left, I had all the parts I needed to restore that engine."

A few years after that, Captain Eddie Sevier passed away. A three-day auction was held to dispose of his shop's contents. Day remembers that auction with a kind of awe. "It took 90 pickup loads of stuff to clear it out," he said. "There were barrels full of carburetors and barrels full of timers and barrels full of other things. The place had been packed to the ceiling."

The only difference between Captain Sevier's place and Dick Day's basement is size.

Because the number of old engines waiting to be found is finite, there is competition among collectors for certain pieces. Once acquired, a particular engine may be kept secret by its owner. Chris Hawes, certainly, has engines he doesn't want others to know about. Dick Day's search for particular Palmer models is unstinting, for over the years he has developed a special affection for that company's products. But others share that affection for Palmer engines. It was one of them, an Eastern Shore enthusiast named Doug Edwards, who acquired a big four-cylinder Palmer in which Day also had been interested.

Doug Edwards is a burly, capable man who talks about his engines almost as if they were members of the family. He speaks in a thoughtful, pleasant drawl in which "retired" sounds like "retarred." The Palmer four, circa 1948, sits out in Edwards' yard under a securely fastened tarp. Since it may be years before he gets around to restoring the engine, he has filled it with over 15 gallons of oil. "This engine," he said, "doesn't have a crack in her and isn't going to get one, either, now that it's retarred."

Edwards' interest in old engines began sometime around 1977. Although he

Doug Edwards in his laundry room.

is not interested in gathering a large collection, he would primarily like to have one-, two-, three-, and four-cylinder T-head Palmers. He already owns several other old machines, each acquired on the limited budget he allows his hobby. Most of them reside on the floor in the family laundry room, next to the red and yellow plastic clothes baskets. Opposite the washing machine and dryer is a little worktable that holds a modest assortment of tools and parts. Work proceeds infinitely slowly in these surroundings. Edwards has been at work for two years on a pair of four-cycle one-lungers, a Falcon and a Palmer YT. His most beautiful engine is a 7 hp Palmer T-header made in 1946. It is installed in an old Chesapeake workboat that Edwards restored over a 1½-year period with the same loving attention and skill he lavishes upon his engines. The boat is a local type, a Hooper Island Drake Tail, a 24-foot dart of a vessel with a 6½-foot beam and two-foot draft.

"I saw this boat over on Bales Creek, near the Choptank," he said, "and right away, I told the wife that I had to have it." There was a sump pump in the bottom of the boat and if it were ever unplugged, the whole rig would have sunk at the dock. The Drake Tail's owner was planning to move to Ohio and was ready to sell the old leaker. He had planned to restore it himself but somehow had never gotten around to it.

Edwards brought the boat home, turned her over, and began the rebuild.

Doug Edwards and his 7 hp Palmer four-cycle.

Every night after he was finished in his boatyard, he worked on it, laboring for about 150 hours each month until the boat was done. By then, she had only one original plank left in her, still bearing the pockmarks of the thousands of bushels of oysters that had once rested against it. The boat's peculiar rounded stern presented a special problem to Edwards. He took off each plank and numbered them all. Then he roughed out replacements on his bandsaw and fitted them together using a hatchet, spokeshave, and high-speed sander. He made a coaming of oak laminations so finely glued that one must look closely to realize it is not one piece of wood. "My skill at steam-bending wasn't up to a one-piece coaming," he said apologetically.

When it was last registered, the Drake Tail had been equipped with a six-cylinder Chevrolet engine that turned a 14-inch propeller, but Edwards planned to use his big, one-cylinder, 7 hp Palmer, although it was missing several parts and had a burnt-out exhaust flange. The parts were found and the flange repaired with the aid of an old-timer who is a superb machinist. It was he who set the Palmer's valve timing without even having a service manual to refer to. Finally, the Palmer, resplendent in a coat of shiny gray paint, was installed in the boat. The engine is cooled via a closed system so that antifreeze can be used and thus greatly reduce corrosion damage. It is fitted with a Schebler carburetor, a six-volt battery, and a Model T jump-spark coil.

On the day we met, Edwards was forced to sweat and tinker before the Palmer settled down to run steadily, its thumping bark voiced through an exhaust pipe that emerges, without benefit of a muffler, through the washboard plank on the port side. He says he plans to replace the drive gears for each of the camshafts because they are worn and emit a discernible rattle. Overall, however, this engine's performance has been exemplary. Doug Edwards figured that during the summer of 1981, he put in a couple of hundred hours on the Palmer, and that it missed exactly two beats while delivering, according to his best reckoning, three hours of running for each gallon of gasoline. Whether the engine is running or not, the boat — named *Christina Lynn* after Edwards' daughter — tends to attract admirers. They gather to examine the gleaming varnish of the oak coamings and the neat gray-and-white paintwork. Restored working craft of any sort are rare, and the Drake Tail is so obviously an unusual antique workboat that it seems to attract more attention than highly polished restored speedboats — which, by comparison, are a dime a dozen.

Edwards uses the boat often on pleasant evenings. She is a wonderful crabbing boat, able to run dead straight along a trot line thanks to her long keel. "She doesn't go all over the creek," is what he said.

Among collectors, opinions are mixed about whether it is really desirable to install a restored marine engine in a boat and use it. Some, like Doug Edwards and Jim Doughty, find it entirely acceptable to do so. Others, however, believe it unwise, primarily from an investment viewpoint. "That's a terrible thing to do to a restored engine," said one.

There is nothing new about such debates. They have raged within the old-car world for years. Some antique automobiles spend all their lives in the cleanliness of antiseptic garages and go to occasional meets in enclosed trailer trucks. Others are driven regularly, according to a philosophy that goes something like, "What's the good of having it if you can't play with it?" Few restored marine engines are used anymore, however. Most lead rather sheltered lives, reposing in barns and sheds where the sunlight slants in to illuminate their old iron heads and big flywheels. Sometimes, one or two may be taken out for display at an engine meet, but, for the most part, old marine engines are allowed to rest in peace. They bask in solitude, visited at their owner's whim. Most collectors are satisfied just to have the old machines around, for the engines exist, as much as anything else, as objects to dwell on, as devices for conjuring up private visions of a world that used to be.

Roger Stone keeps his engine collection in two sheds that are surrounded by spruce trees and a vegetable garden. The Stones live at the farthest end of a dirt lane on a secluded point overlooking the sea in Maine. A few of Stone's choicest engines are in his house, where they stand behind doors or in a corner. There is a

Roger Stone, collector.

A tiny Cady 1½ hp,
Canastota, New
York.

beautiful red Palmer two-cycle (circa 1910) in a bedroom, a tiny gray Cady in the living room. The rest of the engines are out in the shingle-roofed sheds, where they stand in silent iron lines against the wooden walls.

Stone began buying old engines while he was a college student in the late 1950s. He became interested in them when he saw a horizontally opposed make-and-break owned by a friend in Massachusetts. Not long afterward, he discovered his first old marine engine. It was lying in a salt hole near his aunt's house in Maine. This machine turned out to be a 7 hp Lathrop, and he acquired it, frozen piston and all, from its owner. He replaced the piston with one he ordered from Acadia, made a new wrist pin, "even though I'm not a machinist," and "doctored up a plate on which to mount a carburetor." He hooked up a battery and a gasoline tank, and the old Lathrop fired at once, albeit with a knock from a worn crank bearing.

After that, Stone kept a lookout for old engines. One day, he looked around and found he had acquired a dozen of the old machines and had leads to several more. He paid as little as $10 for them and as much as $150. Now, he thinks the prices some people ask for old engines are ridiculous — one reason he has not bought an engine for a long time. His current collection includes, besides the Lathrop, the Palmer, and the Cady, a Knox, a Buffalo, a Bridgeport, a Hubbard, a Roberts, a black one-lunger of unknown manufacture, and a four-cycle Winkley once owned by naval architect Fenwick Williams and once used at the Marblehead Transportation Company. "That's the only engine I've got that I'd be afraid to run again," said Stone. "I'd be afraid it might blow up."

Left to right: Bridgeport 4 hp, Hubbard 4 hp, Winkley.

Stone is somewhat rare among collectors today in that he has some personal memories of days when one-lungers were still in use. He grew up in Marblehead during the 1940s, when there were still a couple of fishermen in town who owned and used boats with the old two-cycle engines.

"I can still hear their putt-putt-putt," said Stone. "Somehow, it seems a shame they're all gone."

The attempt to stave off that ultimate loss, that dread day when these old marine engines might really be "all gone" or, worse, forgotten, lies at the core of the collecting hobby. Lacking, at present, is the overriding profit motive that characterizes so much of the old-car world and many other hobbies. Until quite recently, in fact, old marine engines changed hands for relatively little. If they seem expensive now, that is in part because they used to sell for $10 or $20. A few collectors, of course, have been caught up in the investment mentality that lurks within the entire world of antiques. But the overriding sentiment among collectors of old marine engines remains affection, pure and simple, for the machines themselves.

Some revere any old engine at all, but others, like Dick Day and Doug Edwards, are interested primarily in one make, Palmer. Their enthusiasm just grew. Others collect only a particular make of engine for more personal reasons — nostalgia, for instance. Willard Wight lives in Camden, Maine, and owns only three engines. Every one of them is a Knox. He owned an Eagle once, but he sold it.

Wight remembers Knox engines from his boyhood in Camden. Across the street from his father's coalyard and wharf was a company owned by the very man who, before he turned to building subchasers during World War I, had built the Knox engine. That man was Major John Bird. Major Bird was born in Rockland in 1868 and stayed in Maine all his life. During the heyday of Knox production, from about 1902 to 1916, the factory was one of the most robust enterprises in Camden, and Knox engines were popular with fishermen up and down the Maine coast.

When Willard Wight was growing up, the Knox engine was already out of production, but the Major was operating a company he called the Knox Marine Exchange, selling parts for the many Knox engines still in use then, and other equipment. Willard Wight used to get candy from the Major's secretary, a dour-looking but friendly woman whom he knew as Miss Dyer. He used to go over to the Knox Marine Exchange, eat his candy, and look at all the engine parts and the ship's knees gathered by Major Bird from Maine to Nova Scotia and sold to boatbuilders.

"I have," said Wight, "a long association with Knox."

One day in the early 1970s, Wight acquired an old Lawley launch and decided he would install an old Knox engine in it. He found one rather quickly. It was owned by a machinist whom he knew in Union, Maine. The machinist also owned an Eagle one-lunger — a jump-spark machine — and Wight bought them both. As far as Wight knows, the Knox had been used only lightly. The worst thing that ever happened to it, he believes, is that it was drowned in the hurricane of 1938. Whoever owned it, the old machinist told Wight, took the engine right apart and cleaned up everything and oiled it. This was about the time that Wight decided the Knox was too good to put into his launch. He sold the boat to Peter Spectre, now an editor at *WoodenBoat*, and asked boatbuilder Malcolm Brewer to build a bed for the engine. This Brewer did, constructing the bed of stout and brightly varnished oak. For a while, the old Knox was on display in the Camden-Rockport Historical Society's museum in Rockport, Maine, but now it is on display in the window of Wight's office on Bay View Street.

Willard Wight did not own his Eagle engine for long before he discovered he did not really want it. "I was interested in Knox because of my memories of them," he said, "because of the local history of it. The Eagle had no meaning for me." He sold the Eagle and began seeking more Knox engines. He did not know of Roger Stone's engine, but he did find a handsome green Knox, circa 1905. It had been bought at auction, together with an old airplane engine, by a man who no longer wanted either. Wight also found a red-painted Knox that came from a Michigan collector who wanted to see it in a good home. Wight says he knows of perhaps a half-dozen other Knox engines. Given the chance, he would try to get them all.

Because he wants to share his interest with the public, Willard Wight displays his engines in his store window, where they elicit much curiosity and speculation on the part of passersby. Sometimes, he will invite one of the curious to turn the black machine over. He often leaves the priming cup closed during this exercise — making the flywheel particularly hard to turn — as a practical joke. "Most of 'em just laugh," he said, "but a few get really upset."

Since he is so interested in preserving the Knox name, Wight has gathered as

Willard Wight and his trio of Knox two-cycle engines.

many old catalogs, advertisements, and other paraphernalia as he can. He has, among other things, an original bill of sale. "I'm lucky to have found that," he said. "I swapped a whiskey for it."

He did all this with the image of Major Bird, Miss Dyer, and his own boyhood days always in his thoughts. It was with some excitement that he learned, quite by accident, that the name of Major Bird's last company could be bought from one of the firm's subsequent owners. The company had ceased to exist as anything but a name, but Wight was glad to buy that. He went to work with his best chisels and carved a beautiful sign, a varnished mahogany offering to a world he once knew.

"I'm going to hang this right up on my building," he said, "right across the street from where it used to be."

The sign reads: "Knox Marine Exchange."

One day while we were talking on the telephone, Chris Hawes told me there had been an unhappy little mishap at his house. One of his children had managed to spill a bottle of ink into a drawer where he kept much of his precious engine literature and a number of rare photographs. Some of this material had been ruined, thus diminishing somewhat Hawes's extensive collection of catalogs, books, and pictures. He has several shelves full of early gas-engine books, including an edition of Hiscox.

"That's the rarest," he said. "Everybody would like to get hold of a copy of that."

Every collector of old engines also has his own collection of old literature. Mostly, these are random gatherings of catalogs or such dog-eared shop manuals as might one day help in a tune-up or rebuild. The catalogs are frequently artful things, beautifully printed and illustrated. Some rival in elegance the brochures once issued by the makers of fine automobiles. Sometimes, it is only the existence of these catalogs that indicates a particular engine may ever have been built, or that those who created it had ever lived.

Since the collectors themselves usually revere the old ways and traditional values, the catalogs speak to them of a better world. "You could believe what you read in those days," said one. Taken as a group, old-marine-engine literature addresses itself to fundamental values, of worth per dollar, of honest materials used unstintingly, of long and faithful service. Here are some words from the Guaranty Motor Company brochure written around 1910:

> The Guaranty Motor Company is owned and controlled by two people. Both are practical mechanics and personally supervise all important details. They do their own managing, superintending, and designing and thereby cut out a lot of expense, and bring the customer in contact with the builders of the engine. Their engines will be found as far east as Newfoundland and as far west as British Columbia. Experience is the only practical teacher in the gasoline engine business as the gasoline engine itself is comparatively new. The builders of the Guaranty Engines get their experience in several ways: by experimenting, by scientific research, by the careful perusal of the foremost European and American magazines on the subject, and from six to eight weeks each season spent with the users of their engines on both fresh and salt water. And last but not least is the experience gained from over 40 users of their engines around Hamilton, men owning hulls of every description.

Catalogs like this were the rule rather than the exception. The copy often had a common touch, an ingenuous tone that had a charm all its own. Such catalogs have also become increasingly valuable. Howard Hoelscher is a Boonton, New Jersey, literature dealer who expanded his listings from automobiles to include marine subjects. Once a project supervisor for General Electric in charge of making filaments for 10,000 different lightbulb types, Hoelscher now devotes himself to his literature business. Much of his home is given over to the sorting and storage of thousands of catalogs that he and his wife take with them in a gold-and-white motorhome to shows around the country. Hoelscher first added some old boat catalogs to his offerings at a Massachusetts flea market in 1978. It was all quickly sold and he began buying more, as much as he could find.

"I took the first batch to Hershey," he said, referring to the world's largest automotive flea market, "and the people who saw it there just went bananas. They'd spend hours going through it."

Some were interested in old outboard-motor catalogs. Others sought catalogs from the very early years of the century. After Hershey, Hoelscher divided his boat file into four categories that he labels as follows: "old stuff; newer; newer still; newest." Thus far, he has found most of his sales to be either from the "old stuff" category or else from the 1950s and 1960s. He is not sure whether sales of literature from the 1930s and 1940s are low because he has little to offer from that period or because collectors simply aren't interested at the moment.

Deciding what will be popular at any given time is part of the old literature business. "It's like shooting dice," said Hoelscher, "I go on gut feeling. It's a cyclical thing. Interest in a particular period or particular subject ebbs and flows."

Hoelscher also goes on gut feeling when it comes to pricing his merchandise. "There are some things," he said, "that will sell quite fast and others that I know I'll have for a while. The rare stuff, the stuff you're going to sell for $75 apiece or so, you've got to be careful about. A dealer can't afford to invest too much because his chances of selling it are slim. But when he does sell it, he will make some money."

Howard Hoelscher says he's noticed that while he gets a good price for his really early marine literature, he still has much of it on hand. This does not seem to bother him, however, for he always enjoys browsing through the catalogs himself. Prices on the oldest pieces he has gathered range from about $2 for a Gray lubrication chart, circa 1912, to $75 or $80 for particularly beautiful old catalogs. Some illustrations from one of them, the Ferro catalog, appear elsewhere in this book.

Hoelscher admits that some of the catalogs are expensive, but he and those who buy them view the pieces as both historical documents and art objects. There is always the argument that, just as Chippendale chairs are no longer made, just as Pierce-Arrow automobiles ceased to be manufactured many years ago, they just don't make marine engine catalogs like they used to.

"You open this stuff up," said Hoelscher, "and you have history, right there in front of you."

When we last spoke, he was preparing to send a list of his available material to Chris Hawes. Hawes's appetite for catalogs, ads, books, and magazines could probably keep several literature dealers busy on a full-time basis.

Turning their hobby into a full-time job is something that few old-engine enthusiasts have managed to do. Chris Hawes, with his living room full of engines and his plans for a working museum, has a greater involvement than most of his fellow collectors. Larry Snow, however, has turned his avocation into a vocation. Snow is a former shop teacher who began "playing around with old engines" in the early 1970s. A Californian, his interest focused naturally on machines that once had been built on the West Coast, four-cycle engines that displayed just as much diversity in design and execution as the two-cycles used elsewhere.

"I feel quite fortunate," said Snow, "that we had so many unique engines out here. There's nothing as exciting to watch as a California marine engine. They're full of Mickey Mouse gadgets."

Much of the four-cycles' fascination is created by their intricate valve train with cams, gears, and pushrods. The engines usually possess open crankcases, allowing one to watch the connecting rod as it makes its continual strokes and the crankshaft, which is bathed, more often than not, by a drip-feed or splash-oil system. "California engines," said Snow, "are fun, but they're messy."

As Snow traveled about the Bay area and to old-engine meets in Washington and Oregon, he met many other collectors. A common problem, he learned, was that of reproducing impossible-to-find parts or having to abandon a restoration project that proved beyond a collector's capacity in terms of tools, time, or skill. Most of the old four-cycles that people were finding proved to be in poor condition, with many cracks in their cylinders or water jackets.

"Most people," said Larry Snow, "don't think those cracks are weldable, but I can weld them. I tear the whole water jacket off the cylinder, clean out all the crud, put the jacket back on, and weld it all back together. That's probably why I've got so many engines. Nobody else wants the ones I get."

One day Snow counted up all the engines nobody else had wanted and found he had over 200 of them. By then he was reproducing parts, in a modest way, for other people's engines and, as demand for this service and entire restorations grew, he founded a shop. He named it "The Engine Room." Here Snow brings back to life the old machines that once carried a generation of California fishermen out from port and home again. He has a foundry where he makes sand castings using aluminum, brass, and bronze. He also has a well-equipped machine shop, all the welding equipment he needs, and the tools necessary to fabricate gas, oil, or water tanks from sheet metal.

The Engine Room has yet to see a two-cycle marine engine. Instead, Snow works on the heavy-duty four-cycles whose manufacturers described them in copy like the following, taken from a Hicks Engine Sales Company brochure of about 1920. The Hicks was an overhead valve machine built in San Francisco.

Years of long use have proved Hicks engines economical of fuel and oil. Long life and freedom from vibration are due to the fine balance maintained in engine parts. There is speed variation of 60 to 600 rpm. The Hicks has no equal for ability to idle and then respond promptly to the throttle. For fishing near breakers, treacherous shoals and where sailing is difficult, the Hicks meets every demand.

Hicks was only one of more than two dozen California engine builders. Snow has done his best to learn about them and to collect such literature as he can find. Some of the material he has found was printed by the Standard Gas Engine Company in both Spanish and English. Reproduced on the back cover are two certificates. One, undated, notes that Standard was awarded a Grand Prize at the Panama-Pacific International Exposition. The other is an award preserved at the Alaska Yukon Pacific Exposition in 1909. These fine engines, built in East Oakland, were the product of what Standard described as "the combined mechanical skill and inventive talent and experience of *several* men, most of whom are either Stockholders or Directors of the Company. All of these men have had years of experience in each branch of gas engine building, even before the first STANDARD was built in the year 1901."

Snow said he enjoys studying the old catalogs he finds just about as much as he enjoys fixing up old engines. By now, he possesses more material than most libraries or museums could muster.

There is no telling where you might find an old-gas-engine enthusiast, but sometimes you know that you're getting warm. While I was driving along a road beside Rose Bay, not far from Lunenburg, I saw a man nailing a hand-lettered sign to a telephone pole. "Dory Motor," it said. The road was narrow, but since there had not been any traffic from either direction for several miles, I stopped opposite him.

"What kind of motor is that you're selling?"

He peered down from the ladder he had leaned against the pole. "Make-and-break," he called down, "four-horse." He was a small, slender man, past 60. He wore a gray work shirt and a hat of the variety that usually says "CAT Diesel Power," but it bore no emblem. "Made before you were born, sonny," he added.

I asked him if he thought he might have any trouble selling it. "Well now," he said, "that's hard to say, sonny." He said this softly, as if he had never considered the question before, as if he still thought the buying or selling of a make-and-break one-lunger was an everyday affair. It isn't, not even in Nova Scotia. Because the old fisherman did not seem to have a ready answer, it seemed impolite, somehow, to ask him anything else. I wished him luck and drove on, looking for the house where Carl Tumblin lived.

I had heard about Tumblin from a Rose Bay boatbuilder named Robert Bryan Crockett. "This man is a wonder," Bob Crockett told me. "I have seen more one- and two-cylinder make-and-break and jump-spark engines of all sizes and makes in his shop than anywhere else I can think of. When I walked into his shop and saw a Hartford sitting on his stand, replete with brass cup and valves, I had to have it."

*Hartford make-and-break restored by Carl Tumblin. Owned
by Bob Crockett, Rose Bay Boatworks.*

Crockett possesses the same level-headed and practical approach to things
that characterizes all builders of boats who remain in business for very long.
But the Hartford got him. He bought the green-painted engine on impulse and
promptly installed it in a 21-foot hard-chine catboat he was building on
speculation in one bay of his waterside shop. He had decided to build the boat,
a Brewer-Wallstrom design that he'd seen in *WoodenBoat* magazine, after a
half-hour's thought. Although he had made a good job of the installation, he
was already prepared to replace the Hartford with a modern engine. "It
seemed just right for this particular boat," he said, "but I'm not sure how many
potential buyers will appreciate it."

Bob Crockett met me over at Carl Tumblin's place so that he could introduce
me to the old mechanic. We arrived when Tumblin was halfway through lunch
and waited out in the yard next to his shop. Looking in through the windows, we
could see long workbenches piled high with metal shavings and projects. Many
tools and parts hung from the shop's wooden beams, but, except for an ancient
pair of journal-polishing tongs, I could identify none of them.

After a while, Tumblin, a tall, big-boned man wearing a plaid shirt and
suspenders, came out to greet us. We got acquainted while leaning over the bed
of Tumblin's truck, and he told me that, although he had never been seriously
hurt in all his years in the Lunenburg shops and foundry, he had lost most of his
right thumb to a table saw right here at home. This did not impair his ability to
work, however, and he is often sought out by those needing expert advice or work
on an old engine, or people who just want to hear about how it used to be in the
old days. At some point or other, Tumblin seems to have met everyone in the
Maritimes who manufactured or repaired gasoline engines.

"Chris Hawes was here to my place," said Tumblin. "He came along with an
old engine on the truck. I been to his place two or three times and he has some
mighty fine engines there."

Tumblin's shop is equipped with everything he needs to restore engines, ex-

Carl Tumblin, June 1981.

cept a foundry. All of the machinery is powered by belts, driven by an International Harvester, 4 hp, make-and-break one-lunger, a four-cycle engine that Tumblin believed was about 60 years old.

"I've got what I need here," he said, nodding at the shop. "If cranks are much out of round, more than ten-thousandths, I have the equipment to machine them. We can babbitt bearings, too."

Carl Tumblin began his career with Lunenburg Foundry in the 1920s as an engine fitter and, in the years since then, he has acquired such diverse experience that he can now tell, after an apparently cursory inspection, what an old engine will need if it is to be brought back. There's not much that could surprise him. He's seen water jackets that developed holes after only a few years while others lasted for a generation. He is a modest man who seems to keep quite busy without ever looking for jobs. Old engines have a way of finding him. Lunenburg Foundry routinely recommends him for rebuilds, and collectors — especially in Nova Scotia, where some enthusiasts have 150 or more engines — seek him out. When we last spoke, he told me he had recently acquired six more engines to restore.

"Where did they all come from?" I asked.

"Oh my," said Tumblin. "All over the country. I picked up one here and there." Most of these were International Harvester stationary engines. The

Hartford that Bob Crockett owned was acquired by Tumblin from a relative. "He had bought it second-hand," said Tumblin, "but it stayed for years in a barn in Broad Cove. He said I could buy it if I wanted it." By the time he was through with the Hartford, it had been transformed from an abandoned relic into the beautiful antique that so attracted its present owner.

"Mr. Tumblin," said Bob Crockett, "is a gifted man."

It was late afternoon by the time I left Carl Tumblin's workshop, picking a path past stacks of old parts and a big two-cylinder awaiting restoration on the wooden floor. Tumblin waved good-bye, asking only that I let him know when "the book about old engines" would be ready. After that, I headed back up the road that runs north out of Rose Bay. The bay sparkled an azure blue in the sunlight, and it was easy to see down through the clear water to the kelp and bright pebbles on the bottom.

I stopped when I reached the place where the fisherman had been nailing up his sign earlier that day and was much surprised to find the sign gone. A couple of dories were moored to stakes just off the grassy shore, and in one of them was the little man in the gray shirt. I got out of my rented car and pointed up to where the "Dory Motor" sign had been.

The fisherman squinted in the afternoon sunlight. "I decided to keep her, sonny," he called.

"How come?"

He paused for a minute and appeared to consider. "Did you want to buy her?"

"No, I was just interested, that's all."

"Well," he shrugged, "I just decided. That's all."

We waved at each other and I got back into the car. Then I heard the one-lunger start up. The sound came across the water. Pock-pock-pock-pock-pock. The old-timer seemed entirely content as he steered his dory steadily out into the bay, just as he had done for years.

It is my experience that it is rather more difficult to recapture directness and simplicity than to advance in the direction of ever more sophistication and complexity. Any third-rate engineer or researcher can increase complexity; but it takes a certain flair of real insight to make things simple again.

E.F. Schumacher, *Small is Beautiful*

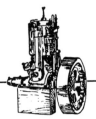

8

The Catboat *Mabel Hawker*

ONE DAY IN AUGUST 1981, I drove to Maryland's Eastern Shore. I went to see a man about his boat. We had been trying to meet ever since May, but delays kept occurring. Once, the boat owner had to go to Idaho to give a lecture. Another time, important family matters prevented our meeting. Finally, in August, arrangements were worked out. I headed south to Maryland, past the fields where DeKalb corn grew higher than a car and the road darted and swept through tiny towns and past a scattering of old houses. At an old general store across from a cornfield, I turned off, slowly passing the Rock Hill Yacht Club and finally arriving at the boatyard.

The vessel I had driven to see was not unusual merely because it had a make-and-break engine. It was unusual because neither the engine nor the boat was a restored antique. The boat had been built in 1973 and, with the owner's approval, the make-and-break engine — a new 4 hp Atlantic — had been installed instead of a more modern gas or diesel machine. Anybody who would have such an engine in his new boat, I thought, must have a story to tell.

By 9:30 a.m., the temperature was already near 80 degrees. Gossamer jellyfish drifted in the water like clouds; blue crabs paddled around the pilings,

Craig Humphrey and Mabel Hawker.

and out on Davis Creek, there was no wind. I spotted the approaching boat — moving silently in the glassy water — when she was still a quarter-mile out. It was easy to make out her distinctive thick mast and a sheer that curved upward in a noble sweep. It was the 18-foot Fenwick Williams-designed cat-boat *Mabel Hawker,* and a family of four was aboard. In a couple of minutes, they reached shore.

"Hi," the skipper said and waved. "Craig Humphrey."

They glided past, the two-cycle's clatter muted by a tight-fitting engine box and an underwater exhaust. I ran to take a dockline and meet Craig

Humphrey, the very man who had no particular interest in old marine engines but who had, for eight years, owned the last example of the type still in production. He used it every weekend, in fact. It seemed strange that a man would do such a thing. It was as if a person were to choose a Model T Ford when he could have ordered a modern automobile.

Mabel Hawker swung 'round in a circle and headed in. I wondered whether Humphrey would reverse her "on the switch" just as the two-ton cat seemed about to demolish the docks, but he did not. He cut the engine and coasted safely in. His wife, Catherine, tied up at the bow, and I took the stern line. As soon as the boat was secured, Humphrey reached down and lifted the engine's hatch.

"Well," he said, "there's what you've come to see."

When he was growing up in the Midwest, Craig Humphrey decided he wanted to sail as a hobby when he was old enough to do so. Later, when he entered graduate school at Brown University to pursue his doctorate in sociology, he used part of a student loan to buy a Lightning, which he kept in a marina. In a slip nearby, a catboat was moored.

"That looked like what a boat should be like," Humphrey said. "I had no idea of how to sail one, but I liked the looks. People who own wooden boats, that's the kind of rationale that makes them do things." He thinks, but is not sure, that the catboat was a Williams design.

In 1967, Humphrey began to look around seriously for a cruising boat to replace the Lightning. He joined the Catboat Association and started studying plans in old magazines. He discovered an old *Rudder* book of plans that included a 19½-foot catboat designed by George Skinner. He liked the boat's looks and began searching for a builder.

"When I learned it was still possible to have a wooden boat built," he remembered, "I asked around and learned of several builders who might do a good job." The bid he received from Roy Blaney in Boothbay, Maine, was the most reasonable. Because it was substantially less than the other bids, however, he became suspicious. He asked all the questions he could think of, but no matter what he asked, the answers always seemed to affirm Blaney's skill. He decided to have Blaney build the boat. But not long afterward, Craig Humphrey received a telephone call from a fellow member of the Catboat Association. The caller suggested that Humphrey ask Blaney to build a Williams-designed catboat rather than the boat he had chosen.

"I asked him why," Humphrey said, "and he answered that it was always best to pick a boat whose designer was fairly well known. I had heard of Fenwick Williams and knew he had worked for John Alden. The idea had a lot of merit." As it turned out, Roy Blaney preferred to build the Williams boat because of the designer's reputation. "That was fine with me," said Humphrey.

He studied the plans for the 18-footer that Williams had drawn in the summer of 1931. The two men began corresponding, and in one letter, Williams wrote: "Wouldn't it be fun to have a small, one-cylinder make-and-break engine in this boat? Of course, in this day and age, people would not tolerate the inconveniences. They want electric start and generators."

But Roy Blaney didn't think electric starting and a generator were so important, and the more Humphrey thought about the matter, the more he liked Williams' suggestion. He talked it over with Roy Blaney, who told him that, if not exactly fun, the make-and-break was desirable for its simplicity. He had already installed several Atlantics in boats he had built, and when he began buying the wood and fittings for the catboat, he suggested that Humphrey buy a 4 hp Atlantic as well.

"He said it was simple," recalled Humphrey, "and that was the best thing it had going for it as far as I was concerned. Because I knew next to nothing about wooden boats except that I liked them, and I knew nothing at all about engines. I'd even sold the outboard engine off the Lightning I owned. So I really needed something simple. Roy liked the engine, Williams said it was a good idea, and I kept hearing that beautiful word — simple."

There was price to consider, too. In 1969, the Atlantic could be had for a mere $400. Humphrey pondered the engine question. He didn't need a generator because he planned to use only kerosene lamps. He remembered an old man from Nova Scotia on Narragansett Bay who had built an H-28 in his backyard. One day the old man told him that a make-and-break engine was the best engine a man could buy. The only problem, the Nova Scotian had said, was that the engines turned big propellers and thus might slow down a sailboat. But that didn't seem important to Humphrey compared with all the advantages. One day he called Roy Blaney and told him to go ahead and buy the one-lunger.

When he contracted for the catboat, Humphrey expected it to be delivered in 1971 or 1972, at the latest. The boat was not completed until October 1973, some five years after he had begun his quest. Because of the delay and the season, the vessel was shipped to Maryland by truck. Humphrey did not have the time to sail his boat home. He had begun a career as a Penn State professor and was content merely to have his new boat at last, safe in a berth. "I believed she would have made it OK," Humphrey said, "because Roy said she would and because other 18-footers of this design have proved very seaworthy."

There was a launching party and the catboat was duly christened. Mabel Hawker is Humphrey's grandmother, who, when she was young, raced sailboats on Lake Macatawa, near her home in Michigan. She was happy to have a boat named for her and proved a critical observer of her grandson's maintenance techniques.

Much to his disappointment and frustration, Humphrey found that his new boat's engine, despite its reputation for simplicity, was not so simple after all. The Atlantic presented him with many questions that he could not answer. The owner's manual was not particularly helpful and neither was the staff at the fancy marina where *Mabel Hawker* was first kept. One cigar-smoking mechanic carefully inspected the Atlantic and said it reminded him of a story he'd heard about two old watermen and a skiff powered by a one-lunger. "One cold morning," he told Humphrey, "they couldn't start their engine and primed it with a lot of ether. The next time they tried to start her, they blew the piston right down through the base and the keel of the boat. That sank their skiff, right there in the Sassafras River!" He puffed on his cigar and gave Humphrey an enigmatic smile.

"They would have jumped in blindfolded to repair a diesel," Humphrey said, "but when they saw that engine in *Mabel Hawker*, they couldn't wait to run the other way." It was a confrontation between the modern mind and an antiquated technology. The one-lunger won every time. Humphrey was left to labor and curse.

His story reminded me of Farley Mowat's *The Boat Who Wouldn't Float*. This is what the Canadian author had to say about the old two-cycle marine engine he'd once owned: "According to mythology, the virtue of these engines lies in the fact that they are simple and reliable. Although this myth is widely believed, I am able to report that it is completely untrue. These engines are, in fact, vindictive, debased, black-minded ladies of no virtue and any non-Newfoundlander who goes shipmates with one is either a fool or a masochist, and is likely both." Mowat's engine was a "seven horsepower, single-cylinder, make-and-break, gasoline monster built in the 1920s from an original design conceived somewhere near the end of the last century. She was massive beyond belief and intractable beyond bearing."

Craig Humphrey is neither a fool nor a masochist, and his Atlantic engine is not a particularly bulky thing. It looks rather friendly, at first glance, with its dome-shaped head and little brass priming cup.

"The way I look at it now," he mused, "I've got a simple engine that runs beautifully, and it only took me seven years to learn how to operate it. There's nothing simple about a simple engine if you don't know anything about it and all the mechanics near you are only comfortable with big, modern engines."

His only sound advice came from a local man who had to care for 50 acres of lawn and knew lots about lawnmower engines. "He could give me some hints," remembered Humphrey. "In fact, when he saw my engine, he became so interested that he started collecting old marine engines. Now he's got a fleet of them."

Listening to her husband recall his first days with the little Atlantic, Cathy Humphrey shook her head. "There have been some times aboard this boat," she said.

Humphrey learned his engine's foibles by experience. Not especially mechanically inclined, he discovered that even a small mistake could have a significant impact on the engine's operation. In his early days of ownership, he had been inconsistent in his mixing of oil and gas, never quite keeping the ratios to the manufacturer's recommendation of one pint to four gallons. The result, he found, was an engine that missed badly.

He learned, as did Jim Doughty with his Lozier, that the timing of a make-and-break engine is critical. "It has to break precisely when the piston is at top dead center," Humphrey said of the Atlantic. "But it's hard to get it timed without the right tools and the right wrenches to get to the nut beneath the eccentric rod and another to get to the one on top of it. If they're not tight, then the timing will change after 10 minutes."

When the timing changed, the Atlantic emitted clouds of black smoke. For two seasons, *Mabel Hawker* sailed with her barndoor rudder blackened by carbon. One day, a sailor suggested politely that if the rudder were painted black, Humphrey would not have to worry about the carbon. Instead, Humphrey took his problem to the owner of the boatyard where he had relocated *Mabel Hawker*. That man happened to be old-engine enthusiast Doug Edwards. One day, Edwards went down to *Mabel Hawker* and adjusted and tightened the timer as it had never been tightened before. The engine has been running like a clock ever since.

In laying up the Atlantic after its first season, Humphrey did as he'd been told and poured a generous quantity of oil into the cylinder. That spring, in the excitement of commissioning the boat, the oil was forgotten until a molten geyser shot out of the cylinder when the engine was turned over. This appeared so alarming that Humphrey telephoned Lunenburg Foundry to ask if he'd done something wrong. But the salesman he spoke with did not understand make-and-break engines and connected him with an old foundryman.

"The old man knew the engine," remembered Humphrey. "The problem was that he was deaf." With the salesman yelling into the old man's ear and the old man yelling back, it was decided that Humphrey should remove the inspection plate on the base and mop out all the oil.

That first season, Humphrey also learned that the engine's battery needed to be fresh. He found this out at about the same time he learned that 7½-volt batteries were about as rare as make-and-break marine engines. "I started using 6-volt batteries or getting those 1½-volt batteries you use for model-airplane engines, but I wired them together," he said. "I knew they wired batteries together in the old days. Still, I always try for one of these 7½-volt dry cells because they are what Roy Blaney said he liked."

By the time he had sorted out the battery situation, the vagaries of his engine's timing, and its demanding diet of a precise oil/gas ratio, Craig Humphrey had learned most of what he needed to know about operating and maintaining his one-lunger. Through it all, he had a lingering sense of betrayal, because the owner's manual was of comparatively little assistance. He thought about the part that warned you not to "crank your head off" and the fact that the pictures were not of his particular model — they were close, but not exact. "That is the sort of thing," he said, "that undermines the naive romantic who would like to operate these things. In the end, you find the engine is as efficient and reliable as they say, but getting to that point, surmounting the obstacles when you start from scratch, is really tough." He leaned back against *Mabel Hawker*'s comfortable cockpit coaming and shook his head. He had the look of a man who has been through something, survived, and learned.

A little before noon, we decided to "go out and run the engine." I hoped the Atlantic was in a better mood than Jim Doughty's Lozier had been, especially since the day was by now very hot. Down the docks, the soda machine was getting as much action as a one-armed bandit in Atlantic City. Humphrey reached down into the engine space.

"I'm closing the choke now," he said, "opening the throttle a bit and adjusting the timing lever. Since she's already been running today, she should start right up." He said this forcefully, as if the engine might be listening.

He next opened the priming cup and filled it with some gasoline from a little oilcan, then closed the ignition switch, put both hands on the flywheel, and rotated it a couple of times until the engine gave a healthy-sounding whistle through the priming cup. Then he closed the cup, turned on the fuel tank tap, and reached down to grasp the starting knob. He pulled the flywheel up hard against compression. The engine did not start. He tried twice more without success. Then he looked critically at the knife switch mounted on the inside edge of the engine box. He opened and shut it a few times.

"This switch," he said, "has corrosion in it."

He gave the flywheel a mighty pull and the engine started. "Backward," he said, "it's going backward." *Mabel Hawker* surged astern and came up short against her bow line. Humphrey cut the switch and the engine stopped.

"Sometimes," he said, "when you pull too hard on the flywheel, that will

Craig Humphrey prepares his Atlantic engine for starting.

happen." He wiped his forehead with a red handkerchief and reached down again to the starting handle.

This time, the engine started in the proper direction, running counterclockwise with a fast-paced pock-pock-pock-pock. We cast off the docklines and turned *Mabel Hawker* out into Davis Creek, cleaving the still water and heading for Langford Bay. Humphrey reached down beneath the engine and, a second later, it speeded up.

"That's full-bore," he said. He had opened the choke and advanced the timing. "It will go faster," he said, "if you close off the water intake and let it get really hot. But it already gets so hot it's unbelievable."

I thought about all the old boat catalogs that emphasized that the engines needed to be installed so as not to scorch any finely finished woodwork. Roy Blaney had wrapped the exhaust manifold of the Atlantic with asbestos and electrician's tape. Both disappeared after the engine had run for 30 or 40 minutes. Humphrey later put a piece of stainless steel on the inside of the engine box by the manifold, but there are still a number of scorch marks in the area.

With the Atlantic churning at full speed, *Mabel Hawker* seemed to make about three knots, perhaps slightly better. Because he seldom uses the engine for anything but docking and powering through a calm, Humphrey finds this performance entirely acceptable. He said the engine's power becomes marginal when pushing *Mabel Hawker*'s ample bulk into a 15- or 20-knot breeze, especially if the tide is foul. What he did not bargain for was the drag created by the 14-inch by 16-inch three-bladed propeller. He remembered the old man in Rhode Island and his admonition about one-lungers' big props only after he noticed that *Mabel Hawker* was slower than most other boats unless running before a strong breeze.

"The propeller," he said, "is like dragging an ash can or bushel basket." He has been contemplating using a two-bladed prop but has not yet done so. Fenwick Williams believes such a prop might work reasonably well.

With the engine hatch cover in place, the Atlantic's noise was not objectionable. It thumped steadily away beneath the wooden cockpit sole, not missing a beat, and once we got out to Langford Bay, we turned to starboard and drove for the point where the bay meets the broad Chester River. We decided to raise sail. While I took the helm, Humphrey took up on *Mabel Hawker*'s topping lift. Then he grasped the throat and peak halyards and the gaff ascended the mast smoothly. When the sail was peaked up, we steered into the wind; Humphrey cut the ignition switch and shut off the gas flow from the gravity-feed tank that was mounted beneath the port cockpit seat.

We began sailing, beating into what was perhaps three knots of wind. I was still at the tiller when Humphrey suggested it was time to tack, since we were "coming right up to a one-foot part." I noticed that the board was still raised and assumed *Mabel Hawker* was often tacked with it in that position. So I trimmed the sheet and put the tiller down, and she came 'round into the eye of the wind — then hesitated there to think about it. I sculled two or three times with the barndoor rudder and around we came, falling off to port.

"Oh," said Humphrey, "the board should have been down." He lowered it and we immediately pointed higher. He settled back as we tacked gently down the bay. He said he thought the propeller must knock a full knot, if not more, off the boat's speed, and as soon as he said that, it seemed as if I could feel the big wheel holding us back. Still, for the sort of sailing the Humphreys most enjoy, gunkholing the many creeks and rivers of the Eastern Shore, the excess drag is not important. After a while, we came about and headed home, running before a breeze that soon disappeared entirely. The sun beat down on *Mabel Hawker* and the sail hung in wrinkles.

"Ah," said Cathy Humphrey, "the Chesapeake in August."

It was time to start the engine again. This time the Atlantic fired right up, and we lowered the sail as we powered into the marina. Humphrey cut the switch as we neared his slip, and we coasted up to a piling and pulled *Mabel Hawker* into her berth. Mrs. Humphrey went below and prepared a lunch of sliced cheese, sausage, and crackers, which we ate in the cabin, with the boat's nine-foot beam accommodating us rather easily. When we finished eating, Humphrey suggested that it was now my turn to "run the engine."

He removed the hatch cover and I stared down at the machine that had originally been gray, but that Humphrey had refinished in red, using Rustoleum primer and Sherwin Williams house paint, "the same that I used on our front door."

Because I have always had an affection for old machinery of all sorts, I have always thought that it had a certain affection for me, too. But now, as I stared at the red Atlantic, I began to have a disturbing sense that when it felt a stranger's hand it might not react at all. It began to look like the most inanimate engine I had ever seen.

Humphrey took me through the starting drill. Close the choke, retard the timing. Fuel off. Open the priming cup. Prime and turn the engine over a couple of times with the ignition switch closed. When the prime hisses in a crisp gasp — as opposed to a dull gasp — shut the priming tap. Turn on the fuel tank tap. I did all these things and then it was time to crank her over. I began to kneel down over the engine and reach down for the flywheel. There didn't seem to be enough room for my feet, so I stood up and reconsidered.

"Try kneeling on the companionway," Humphrey suggested.

I did this and reached down again for the flywheel's starting handle. If you grasp it with your fingertips, it can be pulled out against the tension of the spring, but it pays to have moderately strong fingertips. I pulled the four-inch handle all the way out. At first I tried to grip it as one holds an automobile's starting crank, with the thumb not encircling the handle in case of a backfire. Failing to get a good grip that way, I wrapped my hand around the handle, pulling the flywheel around clockwise, up against compression. I tried to do this smartly, but I managed to do it sluggishly. The Atlantic acted as though I did not exist.

"I guess it takes more muscle," I said. "Wait until next time, flywheel," I thought.

"Well," said Craig Humphrey, "it's just as much a certain touch, really, as muscle. You'll get the hang of it."

I reached back down and pulled the handle out again and heaved on the flywheel. This time I felt more coordinated and the piston came up smartly against compression. Nothing happened. I did this three more times and then reached for the red handkerchief. I primed the engine again with gas from the little oilcan and cranked a few more times, in vain. Humphrey stood above, on the dock.

"I think he's glad," said his wife, "to see somebody else down here struggling with it for a change."

"Wait," said Humphrey. He stepped down into the cockpit and unclipped the wire that went from the ignition switch to the igniter. He brushed it against the cylinder but there was no spark. "It's the switch," he said. "I'll have to take it apart soon and clean it. It gets corrosion inside it." He rapidly opened the switch and closed it a few times and burnished all the metal parts he could reach. "OK," he said, "try her now."

I turned the fuel tap back on, reached down, and pulled out the starting handle, but I let it snap back in when I remembered I had to prime the cylinder. Then I reached for the handle again. I was about ready for a soda or some tea that Cathy Humphrey kept in an ice-filled cooler. I thought, "I hope you're ready to go now that your ignition switch has been cleaned," and gave a hard pull upward. The Atlantic was running. Pock-pock-pock-pock-pock. I thought I had never heard a more gratifying sound. *Mabel Hawker* surged against her docklines, and Cathy Humphrey lunged to raise the dinghy painter away from the spinning propeller.

I started the engine a couple of more times, and then, just when I was feeling satisfied, Humphrey suggested it was time to "reverse her on the switch." At once, my satisfaction turned to doubt. I thought of the story an old-timer had told me about a Greenport, New York, lobsterman. Someone had bet him $50 that he couldn't run his boat up to the dock and then reverse her 50 times in a row. The lobsterman accepted the challenge and lost the bet on his 49th try, or 50th — the old-timer couldn't remember which. I had very dubious feelings about my ability to reverse the Atlantic even once.

Humphrey called down, "It's all a matter of timing."

"What do I watch?"

"You don't watch," he said. "You listen and sense."

I looked at the spinning red flywheel and at the knife switch. The igniter was hammering away in a blur. I didn't feel I had the sense of things. I reached down and opened the switch, hoping somehow to trick the engine before it knew what I was doing, and then quickly closed it. The Atlantic thumped on, unconcerned. Forward. I opened the switch and held it for an instant longer, then closed it. Forward.

"I don't think I have the sense of it," I said.

"I think you're holding it open for too brief a time," said Humphrey.

I feared that if I held the switch open any longer, the engine would stall and I'd have to reach down for the starting handle and begin all over again. I opened the switch wide, counted "one-two-three," and closed it. Backward! *Mabel Hawker* was now pulling against her bow lines. I felt like I'd just passed a driver's test. I opened the switch again, counted "one-two-three," and closed it. Forward. Then I reversed her again. After that, I decided to quit while I was ahead and shut the engine down. I had more respect than ever for those who had once used these machines on a daily basis, for the lobsterman who had lost out on $50, and for Craig Humphrey.

Like most of those involved with make-and-break engines, Craig Humphrey tends to dwell upon not only what they *are* but also what they have come to *represent.* As a sociologist, he has, perhaps, given the matter even more thought than others. He is coauthor of a book titled *Environment, Energy and Society.* Among other things, it explores the idea that many of the problems facing the world today are the result of highly developed technologies and their effects upon people and upon the earth's resources. He sees the one-lunger as an intermediate technology that once had an important place in the world, only to be passed by as new engines were developed and boatmen became more prosperous.

"It is very difficult," said Humphrey, "to show students what intermediate technology is, anymore. A make-and-break marine engine is one of the few good examples left. Of course, it is entirely foreign to almost all of them, but they can readily understand what it once meant — and, for that matter, what it could still mean today in developing nations. There, it would offer an understandable, approachable technology, not one so complex as to victimize its operator completely if something broke down."

This idea — that a machine's simplicity and ultimate function may be virtues that have been forgotten in a rush to more complex, larger models — was set forth some years ago by a well-known British economist in a celebrated book pointing out the dangers inherent in a society's move to increasingly complicated and expensive machines. E.F. Schumacher called his book *Small is Beautiful,* and this is what he said characterized an intermediate technology:

> The equipment would be fairly simple and therefore understandable, suitable for maintenance and repair on the spot. Simple equipment is normally far less dependent on raw materials of great purity or exact specifications than highly sophisticated equipment. Men are more easily trained; supervision, control, and organization are simpler; and there is far less vulnerability to unforeseen difficulties.

Schumacher may not have been thinking about make-and-break marine engines when he wrote his book, but they are, in part, what his book is about. They represent what he would call a "technology with a human face."

Not long after we met, Craig Humphrey regretfully sold *Mabel Hawker* so that his growing family would have more room aboard a big Crocker-designed cutter. The catboat was bought not by an old engine buff, but by John and Donna Cawthorne, who are little older than some of Humphrey's students. They wanted a Williams-designed catboat, and the Atlantic engine only made

Mabel Hawker that much more desirable. John Cawthorne is a skilled mechanic, and with Humphrey's instruction, he came to grips almost immediately with the one-lunger.

It does not matter to John Cawthorne that his boat's engine was designed many years before he was born. He works on complicated modern engines everyday, and he quickly grasped the make-and-break's essential virtues. He has not read E.F. Schumacher's book, but he understands instinctively what a human technology is and what it means. The little Atlantic looked a lot more beautiful to him than any modern, more efficient engine possibly could. It spoke to him about a way of life — a simpler and more self-sufficient way than most people these days can achieve. And that, perhaps more than anything, is the engine's real beauty.

Appendixes

Appendix A

Marine Engine Builders

The following list of marine engine builders in the United States and Canada was compiled with the assistance of Seattle collector Dudley Davidson. The companies listed span a period of some 65 years, from about 1885 until the mid-1950s. Their foundings and failures went, for the most part, unnoticed and unrecorded outside their immediate vicinity. Thus, dates have not been included. Doubtless, there are companies that failed to surface during research. Others were not included because only an engine name and nothing more could be found.

The list includes makers of two-cycle, four-cycle, and diesel engines. The company name is given as it was most commonly listed. It is followed by the city and state where the engine was built and the engine's trade name. An asterisk (*) signifies those companies that built only four-cycle machines. A double asterisk (**) denotes makers of diesel or semidiesel types. The absence of an asterisk indicates that a company primarily was making two-cycle engines.

Anyone having further information about the companies listed, or about other companies, is urged to write the publisher so that the material may be considered for inclusion in a subsequent edition.

UNITED STATES

A.A. Adams & Company
Providence, RI
Trade name: Adams

Able Engine Company
Peekskill, NY
Trade name: Able

*Acme Engine Company
San Francisco, CA
Trade name: Acme

Aerothrust Engine Company
La Porte, IN
Trade name: Aerothrust

*A.G. Hebgen
San Francisco, CA
Trade name: Liberty Kid

A.J. Houle Motor Works
Holyoke, MA
Trade name: (?) (two- and four-cycle)

Alamo. *See* Grand Rapids Engine
   Company

Alexander & Co.
Chicago, IL
Trade name: Humming Bird

All-in-One. *See* Cleveland Marine Motor

Allison Engineering Co.
Indiana
Trade name: Allison

American Engine Co.
Detroit, MI
Trade name: American

American Gasolene Motor Co.
Baldwinsville, NY
Trade name: American

American Motor Company
Eau Claire, WI
Trade name: American

American Motor Works
New York, NY
Trade name: American

Amphion Marine Engines
Milwaukee, WI
Trade name: Amphion

Anderson Engine Company
Shelbyville, IL
Trade name: Anderson

Ardmore. *See* Henry Keidel & Co.

Arrow. *See* Pausin Engineering
   Company

Arrow Motor & Machine Co.
Newark, NJ
Trade name: Waterman

*Astoria Iron Works
Astoria, OR
Trade name: Troger-Fox

Atlantic Company
Amesbury, MA
Trade name: Atlantic

*Atlas (later Atlas-Imperial)
San Francisco, CA
Trade name: Atlas

Auger Engine Company
Minneapolis, MN
Trade name: Auger

*August Mietz Corporation
New York, NY
Trade name: Mietz

Auto Engine Works
St. Paul, MN
Trade name: The Capitol

*Automatic Machine Company
Bridgeport, CT
Trade name: The Automatic

Auto Vim. *See* Buick Motor Company

Baird & Henselwood Co.
Detroit, MI
Trade name: Yale

**Baltimore Oil Engine Company
Baltimore, MD
Trade name: B.O.E.C.

Barber Brothers
Syracuse, NY
Trade name: Barber

Barker Motor Company
Norwalk, CT
Trade name: Barker

Beaver Mfg. Co.
Milwaukee, WI
Trade name: Beaver

Benton. *See* Thomas P. Benton & Son

B.F. Browne Gas Engine Company
Syracuse (later Schenectady), NY
Trade name: Brownie

B.F. Sturtevant Company
Boston, MA
Trade name: Sturtevant

Bicknell Mfg. & Supply Co.
Janesville, WI
Trade name: Bicknell

*Bielfuss Motor Co.
Lansing, MI
Trade name: Bielfuss

**Blanchard Machine Co.
Cambridge, MA
Trade name: Blanchard

Blomstrom. *See* C.H. Blomstrom Motor
Co.

Blount Engineering Co.
Boston, MA
Trade name: Blount

B.O.E.C. *See* Baltimore Oil Engine
Company

**Bolinder Company
New York, NY
Trade name: Bolinder

Boothbay Engine Company
Boothbay, ME
Trade name: Boothbay

Borden & Selleck Co.
Chicago, IL
Trade name: (?)

*Boyer
Oakland, CA
Trade name: Boyer

Brass Jacket. *See* Grant-Ferris Co.

*Brennan Motor Mfg. Co.
Syracuse, NY
Trade name: Brennan Standard

Bridgeport Motor Co.
Bridgeport, CT
Trade name: Bridgeport

Brownback. *See* H.L. Brownback Co.

Brownell-Trebert Co.
Rochester, NY
Trade name: (?)

Brownie. *See* B.F. Browne Gas Engine
Company

Brown-Talbot Machinery Company
Salem, MA
Trade name: Brown-Talbot

Brushmarine. *See* Wilpen Company:
Detroit

*Buda
Harvey, IL
Trade name: Buda

Bud-E. *See* Carlyle Johnson Machine
Co.

*Buffalo Gasolene Motor Company
Buffalo, NY
Trade name: Buffalo

Buick Motor Company
Flint, MI
Trade name: Auto Vim (?) (two- and
four-cycle)

Bull Pup. *See* Fairfield Motor Company

*Burpee and Letson Ltd.
South Bellingham, WA
Trade name: Burpee and Letson

**Busch-Sulzer Bros. Diesel Engine Co.
St. Louis, MO
Trade name: Diesel

Cady. *See* C.N. Cady Co., Inc.

Caille Perfection Motor Company
Detroit, MI
Trade name: Caille

*California Gas Engine Company
San Francisco, CA
Trade name: California

Camden Anchor-Rockland Machine Co.
Rockland, ME
Trade name: Knox

Cameron Marine Car Co.
Beverly, MA
Trade name: (?)

*Campbell Marine Motor Co.
Minneapolis (Wayzata), MN
Trade name: Campbell

Capital. *See* Eld Brothers Company

The Capitol. *See* Auto Engine Works

Carlyle Johnson Machine Co.
Manchester, CT
Trade name: Bud-E

C.F. Sparks Machine Co.
Alton, IL
Trade name: Sparks

*Chandler Dunlop Co.
Seattle, WA
Trade name: Standard Kid (Clift)

*Chas. J. Jager Co.
Boston, MA
Trade name: Jager

*Chas. P. Crouch & Co.
Chicago, IL
Trade name: Junior

Chase Yacht & Engine Company
Providence, RI
Trade name: (?)

C.H. Blomstrom Motor Co.
Detroit, MI
Trade name: Blomstrom

*Chesapeake Engine Co.
Oxford, MD
Trade name: Chesapeake

*Christie Machine Works
San Francisco, CA
Trade name: Christie

Clark Manufacturing Company
Fond du Lac, WI
Trade name: (?)

*Claude Sintz
Grand Rapids, MI
Trade name: Leader

Claus. *See* International Motor
    Company

*Clay Engine Mfg. Co.
Cleveland, OH
Trade name: Honest Clay

Cleveland Marine Motor
Cleveland, OH
Trade name: All-in-One

Clift. *See* Chandler Dunlop Co.;
    Clift Motor Co.

*Clift Motor Co.
Bellingham, WA
Trade name: Clift

*Clifton Motor Works
Cincinnati, OH
Trade name: Clifton

Climax Engineering Company
Clinton, IA
Trade name: Climax

C.M. Giddings
Rockford, IL
Trade name: Sure Go

C.N. Cady Co., Inc.
Canastota, NY
Trade name: Cady

Colchester. *See* Marblehead Machine
    Company

Columbia Engine Company
Detroit, MI
Trade name: (?)

*Commonwealth Motor Company
Chicago, IL
Trade name: Quayle

Consistent. *See* Milwaukee Motor
Mfg. Co.

Consolidated Shipbuilding Corp.
(formerly Gas Engine & Power Co. &
Charles L. Seabury & Co. Cons.)
New York, NY
Trade name: Speedway

The Cook Engine Company
Cambridge, IL
Trade name: Cook

*Corliss
Petaluma (later San Francisco), CA
Trade name: Corliss

Crabber. *See* Loane Engineering
Company

Cragg. *See* Gilmore-Cragg Motor
Manufacturing Company

*Craig Marine Gasolene Engines
New York, NY
Trade name: Craig

Crescent Marine Motor Co.
St. Louis, MO
Trade name: (?)

Cross. *See* M.O. Cross Engine Co.

Crown. *See* Yacht Gas Engine & Launch
Co.

C.T. Wright Engine Company
Greenville, MI
Trade name: Weco

*Cummins Engine Co.
Columbus, IN
Trade name: Cummins

Cushman Motor Company
Lincoln, NE
Trade name: Cushman (two- and four-
cycle)

*Daimler Motor Company
New York, NY
Trade name: Daimler

*Davis Manufacturing Co.
Milwaukee, WI
Trade name: Davis

Dean Manufacturing Company
Newport, KY
Trade name: Fox

*Delaware Marine Motor Co.
Wilmington, DE
Trade name: Delaware (?)

Delong Engine Co.
Webster, NY
Trade name: (?)

Detroit Auto Engine Specialty Company
Detroit, MI
Trade name: Liberty

Detroit Engine Works
Detroit, MI
Trade name: Detroit

Detroit Gas Engine & Machinery Co.
Detroit, MI
Trade name: Major

Diesel. *See* Busch-Sulzer Bros. Diesel
Engine Co.

*Doak Gas Engine Company
San Francisco/Oakland, CA
Trade name: Doak

**Dodge Sales and Engineering Co.
Mishawaka, IN
Trade name: Dodge Heavy Oil Engine

Doman. *See* H.C. Doman Co.; Universal
Products Co.

**Dow Pump and Diesel Engine Company
Alameda, CA
Trade name: Dow

DuBrie Motor Co.
Detroit, MI
Trade name: DuBrie

*Duesenberg
Elizabeth, NJ (previously St. Paul and
Chicago)
Trade name: Duesenberg

Dunn Motor Works
Ogdensburg, NY
Trade name: Dunn (two- and four-cycle)

Dutcher Machine Company
Fulton, NY
Trade name: New Parker

Dutton Engineering
Perth Amboy, NJ
Trade name: Dutton

Eagle. *See* Eagle Company; Progressive
   Manufacturing Company

Eagle Company
Newark, NJ (moved from Torrington,
   CT, ca. 1910)
Trade name: Eagle

Easternhouse Engine Co.
Taunton, MA
Trade name: Pearl

*Eclipse Motor Co.
Mancelona, MI
Trade name: Eclipse

E. Garry Emmons
Swampscott, MA
Trade name: Emmons

Elbridge Engine Company
Rochester, NY
Trade name: Elbridge

*Elco Works
Bayonne, NJ
Trade name: Elco (formerly J.V.B.)

Eld Brothers Company
Augusta, ME
Trade name: Capital

Emerson Motor Company
Alexandria, VA
Trade name: Emerson (two- and four-
   cycle)

Emery. *See* V.J. Emery

Emmons. *See* E. Garry Emmons

**Enterprise Foundry Co.
San Francisco, CA
Trade name: Enterprise

*Enterprise Machine Co.
Minneapolis, MN
Trade name: Westman

Erd Motor Company
Saginaw, MI
Trade name: Erd

Essex. *See* The Essex Engine Co.;
   W.J. Young Machinery Co.

The Essex Engine Co.
Lynn, MA
Trade name: Essex

Evansville Gas Engine Works
Evansville, IN
Trade name: Evansville

Fairbanks Morse & Co.
Chicago, IL
Trade name: Fairbanks Morse (two- and
   four-cycle)

Fairfield Motor Company
Bridgeport, CT
Trade name: Bull Pup

*F.A. Seitz & Co.
Newark, NJ
Trade name: (?)

Fay & Bowen Engine Company
Geneva, NY
Trade name: Fay & Bowen (two- and
   four-cycle)

Ferro Machine & Foundry
Cleveland, OH
Trade name: Ferro

Fifield Brothers
Augusta, ME
Trade name: Fifield

*Fishback Motor Co.
Detroit, MI
Trade name: Fishback

Fisherman. *See* Loane Engineering
   Company

Fisherman-Ford. *See* Loane Engineering
   Company

F.J. Humphreys
Skaneateles, NY
Trade name: Gosper

*F.K. Lord
New York, NY
Trade name: Lord

Foster Motor Company
New Haven, CT
Trade name: Foster

Fox. *See* Dean Manufacturing Company

Fraser Machine & Mfg.
Boston, MA
Trade name: Fraser

Frazer Bros. Co.
Adams, NY
Trade name: Frazer-Adams

*Fred J. Pease Co.
Grand Rapids, MI
Trade name: Pease

*Frisbie Motor Company
Middletown, CT
Trade name: Frisbie

Fulton Engine Company
Erie, PA
Trade name: Fulton

FW. *See* Mechanical Engineers Investing
    Company

F.W. Sherman
Buffalo, NY
Trade name: Sherman

*Gaeth
Cleveland, OH
Trade name: Gaeth

Gaffga. *See* J.S. Gaffga & Co.

*Gas Engine & Power Co.
Morris Heights, NY
Trade name: Speedway

G.B. & S. *See* Golden, Belknap & Swartz
    Co.

General Machinery Company
Bay City, MI
Trade name: Smalley

Genesee Launch & Power Co.
Rochester, NY
Trade name: Genesee

Gierholtt Machinery Company
Marine City, MI
Trade name: Gierholtt

Gile Boat and Engine Company
Ludington, MI
Trade name: Gile

Gilmore-Cragg Motor Manufacturing
    Company
Chicago, IL
Trade name: Cragg

*Gilmore Motor Mfg. Co.
Detroit, MI
Trade name: Gilmore (two- and four-
    cycle)

Globe. *See* Pennsylvania Iron Works
    Company

*Globe Iron Works
Minneapolis, MN
Trade name: White

*Golden, Belknap & Swartz Co.
Detroit, MI
Trade name: G.B. & S.

*Golden Gate
San Francisco, CA
Trade name: Golden Gate

*Gorham Engineering
Alameda, CA
Trade name: Gorham

Goshen Motor Works (later White
    Manufacturing Company)
Goshen, IN
Trade name: Goshen

Gosper. *See* F.J. Humphreys

*Grand Rapids Engine Company
Hillsdale, MI
Trade name: Alamo

*Grand Rapids Yacht & Engine Co.
Grand Rapids, MI
Trade name: Monarch

Grant-Ferris Co.
Troy, NY
Trade name: Brass Jacket

Grasser Motor Company
Toledo, OH
Trade name: Grasser

Gray & Prior Machine Company
Hartford, CT
Trade name: Hartford

Gray Motor Company
Detroit, MI
Trade name: Gray (two- and four-cycle)

*Grimm Mfg. Co.
Buffalo, NY
Trade name: Grimm

*Gulowsen-Grei
Seattle, WA
Trade name: Gulowsen-Grei

*Haase Motor Works
Milwaukee, WI
Trade name: Haase

Hallett Mfg. Co.
Los Angeles, CA
Trade name: Hallett

Hall Gas Engine Manufacturing
Company
Byesville, OH
Trade name: Hall

*Hallin Engine Manufacturing Co.
Tacoma, WA
Trade name: Hallin

*Hall-Scott Motor Car Co.
Berkeley, CA
Trade name: Hall-Scott

H & H. *See* Hurd and Haggin

Hanley Marine. *See* Midland Engine
Works Co.

Harrison. *See* Magann Air Brake Co.

Hartford. *See* Gray & Prior Machine
Company

Harvey Marine Motor Company
Rochester, NY
Trade name: Harvey Motors

*Hazard Motor Mfg. Co.
Rochester, NY
Trade name: Hazard

*H.C. Doman Co.
Oshkosh, WI
Trade name: Doman

Henry Keidel & Co.
Baltimore, MD
Trade name: Ardmore

*Hercules Gas Engine Company
San Francisco, CA
Trade name: Hercules

*Herrmann Engineering Co.
Detroit, MI
Trade name: The Small Aristocrat

*Hess Motors (Algonac Machine & Boat
Works)
Algonac, MI
Trade name: Hess-Mono-Marine

Hettinger Engine Company
Bridgeton, NJ
Trade name: (?) (two- and four-cycle)

*Hicks Engine Sales Company
San Francisco, CA
Trade name: Hicks

Hildreth Mfg. Co.
Lansing, MI
Trade name: Hildreth

**Hill Diesel Engine Company
Lansing, MI
Trade name: Hill

H.J. Leighton
Syracuse, NY
Trade name: Leighton

H.L. Brownback Co.
Norristown, PA
Trade name: Brownback

H.L.F. Trebert
Rochester, NY
Trade name: (?)

*Holiday Engineering Co.
Chicago, IL
Trade name: Holiday

Holmes-Howard Motor Co.
Detroit, MI
Trade name: Holmes (two- and four-
cycle)

Honest Clay. *See* Clay Engine Mfg. Co.

Hoosier Motor Co.
Goshen, IN
Trade name: Hoosier

Howard Motor Company
Philadelphia, PA
Trade name: Howard

*Hub Motor Co.
Boston, MA
Trade name: Hub

Hubbard Motor Company
Middletown, CT
Trade name: Hubbard

*Humboldt Gas Engine
San Francisco, CA
Trade name: Humboldt

Humming Bird. *See* Alexander & Co.

*Hurd and Haggin
New York, NY
Trade name: H & H

**H.W. Sumner
Seattle, WA
Trade name: Sumner

*Imperial Gas Engine Co.
San Francisco, CA
Trade name: Imperial

International Motor Company
Detroit, MI
Trade name: Claus

**International Power Vehicle Co.
Stamford, CT
Trade name: International

International 16. *See* Sutter Bros.

**Jacobsen Gas Engine Co.
Sarasota Springs, NY
Trade name: Jacobsen

Jager. *See* Chas. J. Jager Co.

Jencick Motor Manufacturing Company
Port Chester, NY
Trade name: Jencick

John Stuart Co.
Wollaston, MA
Trade name: Stuart

J.S. Gaffga & Co.
Greenport, NY
Trade name: Gaffga

*Jule Motor Corp.
Syracuse, NY
Trade name: Jule

Junior. *See* Chas. P. Crouch & Co.

J.V.B. *See* Elco Works

J.W. Lathrop & Co.
Mystic, CT
Trade name: Lathrop (two- and four-
cycle)

*Kahlenberg
Two Rivers, WI
Trade name: Kahlenberg

Kennebec. *See* Torrey Roller Bushing
Works

*Kermath Motor Co.
Detroit, MI
Trade name: Kermath

**Kerosene Power Co.
Minneapolis, MN
Trade name: Viking

Knox. *See* Camden Anchor-Rockland
Machine Co.

*Knox Motors Co.
Springfield, MA
Trade name: Knox-Springfield

Knox-Springfield. *See* Knox Motors Co.

L-A. *See* Lockwood-Ash Motor
Company

Lackawanna (?)
Newburgh, NY
Trade name: Lackawanna

Lacy Brothers
Toledo, OH
Trade name: Lacy

*Lamb Boat and Engine Company
Clinton, IA
Trade name: Lamb   .

Lathrop. *See* J.W. Lathrop & Co.

Leader. *See* Claude Sintz

Leary Gasoline Engine Co.
Rochester, NY
Trade name: Leary

Leighton. *See* H.J. Leighton

*Le Roi Company
Milwaukee, WI
Trade name: Le Roi

Liberty. *See* Detroit Auto Engine
   Specialty Company

Liberty Kid. *See* A.G. Hebgen

Little Skipper. *See* Monarch Tool and
   Engine Company

Little Tiger. *See* Northwestern Machine
   Co.

*Livingston & Van Epps
Syracuse, NY
Trade name: Van Epps

Lloyd. *See* Thousand Island Boat &
   Engine Co.

*Loane Engineering Company
Baltimore, MD
Trade names: Fisherman, Crabber,
   Fisherman-Ford

Lockwood-Ash. *See* Lockwood-Ash
   Motor Company; Nadler Foundry &
   Machine Co., Inc.

Lockwood-Ash Motor Company
Jackson, MI
Trade name: L-A

*Loew Manufacturing Co.
Cleveland, OH
Trade name: Loew Victor

London Motor Company
New London, CT
Trade name: Pequot

Lord. *See* F.K. Lord

**Lorimer
Oakland, CA
Trade name: Lorimer

Lozier Motor Company
Plattsburgh, NY (formerly Toledo, OH)
Trade name: Lozier (two- and four-cycle)

*Lycoming Manufacturing Company
Williamsport, PA
Trade name: Lycoming

McDuff
Lakeville, NH
Trade name: McDuff

**M'Intosh & Seymour Corp.
Auburn, NY
Trade name: (?)

M.A.F. *See* Rome Gasoline Engine
   Company

Magann Air Brake Co.
Detroit, MI
Trade name: Harrison

Major. *See* Detroit Gas Engine &
   Machinery Co.

M. & T. *See* Murray & Tregurtha

Manhattan Manufacturing Co.
New York, NY
Trade name: Manhattan

Marblehead Machine Company
Marblehead, MA
Trade name: Colchester

Marine Motor Manufacturing Co.
Boston, MA
Trade name: (?)

Marine Power Co.
Milwaukee, WI
Trade name: (?) (two- and four-cycle)

*Mason Machine Works
Taunton, MA
Trade name: Mason

*Maynard-Adams Engine Co.
Berkeley, CA
Trade name: Maynard-Adams

Meade Gas Engine Co.
Providence, RI
Trade name: Meade

*Mecco Engine Co.
Philadelphia, PA
Trade name: Mecco

*Mechanical Engineers Investing
  Company
Los Angeles, CA
Trade name: FW

*Mechanics Foundry & Machine
Fall River, MA
Trade name: Mechanics

*Mercury Motor Company
Long Island City, NY
Trade name: Mercury

Mianus Motor Works
Mianus/Stamford, CT
Trade name: Mianus

*Michigan Marine Motor Corp.
Detroit, MI
Trade name: Michigan Marine Motor

Mietz. *See* August Mietz Corporation

*Miller Engine Company
Chicago, IL
Trade name: Miller

*Milwaukee Motor Mfg. Co.
Milwaukee, WI
Trade name: Wisconsin (Consistent)

**Missouri Engine Company
St. Louis, MO
Trade name: Missouri

Miss Simplicity. *See* St. Joseph Motor
  Company

M.O. Cross Engine Co.
Detroit, MI
Trade name: Cross

Mohawk. *See* S-R Mfg. Co.

Monarch. *See* Grand Rapids Yacht &
  Engine Co.

Monarch Tool and Engine Company
Cincinnati, OH
Trade name: Little Skipper

Moore. *See* Palmer Moore Co.

*Morristown Boat & Engine Co.
Morristown, NY
Trade name: Morristown

*Morro Motor Company
Morro, CA
Trade name: Morro

Morton Motor Co.
Detroit, MI
Trade name: Morton

Motor de Luxe. *See* Motor Vehicle Power
  Co.

Motor-Go. *See* Nadler Foundry and
  Machine Co., Inc.; Sears, Roebuck &
  Co.

Motor Vehicle Power Co.
Philadelphia, PA
Trade name: Motor de Luxe (two- and
  four-cycle)

*Murray & Tregurtha
Atlantic (later South Boston), MA
Trade name: M. & T.

Nadler Foundry and Machine Co., Inc.
Plaquemine, LA
Trade name: Nadler (engine was former-
  ly known as Sears Motor-Go and
  Lockwood-Ash)

*N and S Engine Co.
Seattle, WA
Trade name: N and S

*National United Service Co.
Detroit, MI
Trade name: Simplex

Neponset Engine & Machine Works
Neponset (Boston), MA
Trade name: Rex

*New Jersey Motors
Keyport, NJ
Trade name: N.J.M.

**New London Ship & Engine Company
Groton, CT
Trade name: NLSECO

New Parker. *See* Dutcher Machine
   Company

**New York Kerosene & Oil Engine Co.
New York, NY
Trade name: New York

*New York Yacht, Launch & Engine Co.
Morris Heights, NY
Trade name: 20th Century

*Niagara Gasoline Motor Company
Buffalo, NY
Trade name: Niagara

*Nichoalds Company
Detroit, MI
Trade name: Nichoalds

N.J.M. *See* New Jersey Motors

NLSECO. *See* New London Ship &
   Engine Company

Northwestern Machine Co.
Detroit, MI
Trade name: Little Tiger

Northwestern Marine Motors
Eau Claire, WI
Trade name: Northwestern

Oakland. *See* Wilpen Company: Detroit

Olympic. *See* Woodhouse Gas Engine
   Co.

Ontario Iron Works
Pulaski, NY
Trade name: Ontario

Oriole. *See* Page Engineering Company

*Oswald Gas Motor Company
San Francisco, CA
Trade name: Oswald

*Otto Gas Engine Works
Philadelphia, PA
Trade name: Otto

**Pacific Diesel Engine Company
Oakland, CA
Trade name: Werkspoor

*Pacific Marine Engine Co.
Seattle, WA
Trade name: Pacific

Page Engineering Company
Baltimore, MD
Trade name: Oriole

Palmer Bros. Engines, Inc.
Cos Cob, CT
Trade name: Palmer (two- and four-
   cycle)

Palmer Moore Co.
Syracuse, NY
Trade name: Moore

Paulsen
Geneva, NY (?)
Trade name: Paulsen

Pausin Engineering Company
Newark, NJ
Trade name: Arrow

Pearl. *See* Easternhouse Engine Co.

Pease. *See* Fred J. Pease Co.

Peerless Marine Engine Co.
Detroit, MI
Trade name: Peerless

*Pennsylvania Iron Works Company
Philadelphia/Eddystone, PA
Trade name: Globe

Penrose Motor, Inc.
Philadelphia, PA
Trade name: Penrose

Pequot. *See* London Motor Company

Perkins Bros.
Havana, IL
Trade name: S & H

Perkins Launch & Motor Co.
Gloucester, MA
Trade name: RoNoMor

Phillips Gas Engine & Motor Works
Chicago, IL
Trade name: Phillips

Pierce-Boutin Motors, Inc.
Merrill, MI
Trade name: Pierce-Boutin

Pilot. *See* Trump Brothers Machine

*Portage Boat & Engine Company
Portage, WI
Trade name: Portage

Potter Manufacturing Co.
New Brunswick, NJ
Trade name: Potter

Powell Engine Corporation
New York, NY
Trade name: Powell

Progressive Manufacturing Company
Torrington, CT
Trade name: Eagle

*Putnam Motor Mfg. Co.
Springfield, MA
Trade name: Putnam

Quale. *See* Commonwealth Motor
    Company

Ralaco. *See* S.M. Jones Company

*Raymond Engineering Co.
Boston, MA
Trade name: RV

*Ray Motor Co.
Detroit, MI
Trade name: Ray

*Red Wing Motor Co.
Red Wing, MN
Trade name: Red Wing

Reeves-Graff. *See* Trenton Engine
    Company

*Regal Gasoline Engine Co.
Coldwater, MI
Trade name: Regal

**Remington Oil Engine Company
Keyport, NJ (also Stamford, CT)
Trade name: Remington

Rex. *See* Neponset Engine & Machine
    Works

Reynolds Motor Company
Detroit, MI
Trade name: Reynolds

*Robbins
San Diego, CA
Trade name: Robbins

Robert S. Hill
Detroit, MI
Trade name: Victor

Roberts Motor Company
Sandusky, OH
Trade name: Roberts

Rochester Gas Engine Company
Rochester, NY
Trade name: Rochester

Rome Gasoline Engine Company
Rome, NY
Trade name: M.A.F.

RoNoMor. *See* Perkins Launch & Motor
    Co.

Royal Engine Company
West Mystic, CT
Trade name: Royal

RV. *See* Raymond Engineering Co.

Sagamore Engine Company
Lynn, MA
Trade name: Sagamore

St. Joseph Motor Company
St. Joseph, MI
Trade name: Miss Simplicity

*St. Lawrence River Motor & Machine
    Co.
Clayton, NY
Trade name: St. Lawrence

*Samson Iron Works
Stockton, CA
Trade name: Samson

S & H. *See* Perkins Bros.

Sarvent Marine Engine Works
Chicago, IL
Trade name: Sarvent

*Scripps Motor Company
Detroit, MI
Trade name: Scripps

Sears Motor-Go. *See* Nadler Foundry
and Machine Co., Inc.; Sears,
Roebuck & Co.

Sears, Roebuck & Co.
Philadelphia, PA (and Chicago)
Trade name: Motor-Go

Seattle Spirit. *See* Woodhouse Gas
Engine Co.

*Seattle Standard Engine Mfg. Co.
Seattle, WA
Trade name: Seattle Standard

Sherman. *See* F.W. Sherman

Shortt. *See* W.M. Ryan Company

Silent Tige. *See* Tige Motor Company

Simplex. *See* National United Service Co.

Sloane Motor Co.
Chicago, IL
Trade name: Sloane

The Small Aristocrat. *See* Herrmann
Engineering Co.

Smalley. *See* General Machinery
Company; Smalley Motor Co.

Smalley Motor Co.
Bay City, MI
Trade name: Smalley

S.M. Jones Company
Toledo, OH
Trade name: Ralaco

**Southwark Foundry & Machine
Philadelphia, PA
Trade name: Southwark-Harris

Sparks. *See* C.F. Sparks Machine Co.

Spaulding Engine Co.
St. Joseph, MI
Trade name: Spaulding

*Speed Changing Pulley Co.
Anderson, IN
Trade name: (?)

Speedway. *See* Consolidated Shipbuilding Corp. and Gas Engine &
Power Co.

Springfield Motor Company
Springfield, MA
Trade name: Springfield

S-R Mfg. Co.
Schenectady, NY
Trade name: Mohawk

*Standard Gas Engine Company
Oakland, CA
Trade name: Standard

Standard Kid (Clift). *See* Chandler
Dunlop Co.

*Standard Motor Construction Co.
Jersey City, NJ
Trade name: Standard

Stanley Company
Boston, MA
Trade name: Stanley

*Stearns Motor Manufacturing Co.
Ludington, MI
Trade name: Stearns Extra Reserve

*Sterling Engine Co.
Buffalo, NY
Trade name: Sterling

Stork Motor Company
Saginaw, MI
Trade name: Stork

Strelinger Marine Engine Co.
Detroit, MI
Trade name: Strelinger

Strickland
Bound Brook, NJ
Trade name: Strickland

Sturtevant. *See* B.F. Sturtevant
Company

Sumner. *See* H.W. Sumner

Superior Motor Works
Jackson, MI
Trade name: Superior

Sure Go. *See* C.M. Giddings

Sutter Bros.
New York, NY
Trade name: International 16

Syracuse Gas Engine Company
Syracuse, NY
Trade name: Syracuse

Terry Engine Co.
Brooklyn, NY
Trade name: Terry

*Thelma Engine Works
Detroit, MI
Trade name: Thelma

Thomas & Grant
Ithaca, NY
Trade name: (?)

Thomas P. Benton & Son
LaCrosse, WI
Trade name: Benton

Thousand Island Boat & Engine Co.
Morristown, NY
Trade name: Lloyd

Thrall Motor Company
Detroit, MI
Trade name: Thrall Refined

*Tige Motor Company
Port Angeles, WA
Trade name: Silent Tige

Toledo. *See* Universal Machine
   Company

Toppan Boat Manufacturing Company
Boston, MA
Trade name: Toppan Reversible

Toquet Motor Company
Saugatuck, CT
Trade name: Toquet

**Torrens & Co.
New York, NY
Trade name: (?)

Torrey Roller Bushing Works
Bath, ME
Trade name: Kennebec

Treibert Engine Co.
Yonkers, NY
Trade name: (?)

**Trenton Engine Company
Trenton, NJ
Trade name: Reeves-Graff

Troger-Fox. *See* Astoria Iron Works

Trojan Launch & Motor Works
Troy, NY
Trade name: Trojan

Trump Brothers Machine
Wilmington, DE
Trade name: Pilot

Truscott Boat Manufacturing Co.
St. Joseph, MI
Trade name: Truscott

Tuttle. *See* Tuttle Motor Company;
   Watertown Motor Company

Tuttle Motor Company
Canastota, NY
Trade name: Tuttle

20th Century. *See* New York Yacht,
   Launch & Engine Co.

*Union Gas Engine Company (result of
   merger of Regan Co. and San Francisco Co.)
Oakland/San Francisco, CA
Trade name: Union

United States Motors Corp. (merged with
   Doman ca. 1928)
Oshkosh, WI
Trade name: US Falcon

Unit Motor Co.
Kansas City, MO
Trade name: Unit Motor

Universal. *See* West Mystic Manufacturing Company; Universal Motor Co.

Universal Machine Company
Toledo, OH
Trade name: Toledo

*Universal Motor Co.
Oshkosh, WI
Trade name: Universal

*Universal Products Co.
Oshkosh, WI
Trade name: Doman

US Falcon. *See* United States Motors
    Corp.

*Van Auken Motor & Machine Works
Bridgeport, CT
Trade name: Van Auken

*Van Blerck Motor Co.
New York, NY (and Detroit)
Trade name: Van Blerck

Van Epps. *See* Livingston & Van Epps

Vanguard Engine Company
Boston, MA
Trade name: Vanguard

Victor. *See* Robert S. Hill

Viking. *See* Kerosene Power Co.; Viking

*Viking
Seattle, WA
Trade name: Viking

Vim Motor Company
Sandusky, OH
Trade name: Vim

*V.J. Emery
Wollaston, MA
Trade name: Emery

*Vulcan Engine Works
Philadelphia, PA
Trade name: Vulcan

**Washington Iron Works
Seattle, WA
Trade name: Washington Estep

Waterman. *See* Arrow Motor & Machine
    Co.; Waterman Marine Motor Works

Waterman Marine Motor Works
Grand Rapids, MI
Trade name: Waterman

Watertown Motor Company
Canastota, NY
Trade name: Tuttle

Watkins Motor Company
Cincinnati, OH
Trade name: Watkins

Weco. *See* C.T. Wright Engine Company

*Wellman-Seaver-Morgan Co.
New York, NY
Trade name: WSM

Werkspoor. *See* Pacific Diesel Engine
    Company

*Western Gas Engine Machine Works
Seattle, WA
Trade name: Western

**Western Machinery Co.
Los Angeles, CA
Trade name: Western

Westman. *See* Enterprise Machine Co.

West Mystic Manufacturing Company
West Mystic, CT
Trade name: Universal

White. *See* Globe Iron Works

Willet Engine & Carburetor Company
Buffalo, NY
Trade name: Willet

*Wilpen Company: Detroit
Detroit, MI
Trade names: Brushmarine, Oakland

Winona Machine & Boat Works
Winona, MN
Trade name: Winona

*Winton Engine Works
Cleveland, OH
Trade name: Winton

Wisconsin. *See* Milwaukee Motor Mfg.
    Co.; Wisconsin Machinery & Mfg. Co.

Wisconsin Machinery & Mfg. Co.
Milwaukee, WI
Trade name: Wisconsin

W.J. Young Machinery Co.
Lynn, MA
Trade name: Essex

W.M. Ryan Company
Chicago, IL
Trade name: Shortt

*Wolverine Motor Works
Bridgeport, CT
Trade name: Wolverine

Wolverine Motor Works
Grand Rapids, MI
Trade name: Wolverine

Wonder Manufacturing Company
Syracuse, NY
Trade name: Wonder

*Woodhouse Gas Engine Co.
Seattle, WA
Trade names: Olympic, Seattle Spirit

**Worthington Pump & Machinery Corp.
New York, NY
Trade name: Worthington Diesel

*Wright Machine Co. (later Gunther-
   Wright)
Owensboro, KY
Trade name: Wright Reliable

Wright Motor Company
Buffalo, NY
Trade name: Wright

WSM. *See* Wellman-Seaver-Morgan

*Yacht Gas Engine & Launch Co.
Philadelphia, PA
Trade name: Crown

Yale. *See* Baird & Henselwood Co.

CANADA

Acadia Gas Engines Limited
Bridgewater, NS
Trade name: Acadia (two- and four-
   cycle)

*Adams
Penetanguishene, ON/Vancouver, BC
Trade name: Adams

Atchison-Bower
Shelburne, NS
Trade name: Etherington

Atlantic. *See* Lunenburg Foundry Ltd.

Barrie. *See* Canada Producer Gas Engine
   Co.

Bruce Stewart & Co., Ltd.
Charlottetown, P.E.I.
Trade name: Imperial

Burrill-Johnson Foundry & Machine
Yarmouth, NS
Trade name: Burrill-Johnson

Canada Producer Gas Engine Co.
Barrie, ON
Trade name: Barrie

Carling Boat Works
Port Carling, ON
Trade name: Carling

Champion Machine and Motor Works,
   Ltd.
St. John's, NF
Trade name: Champion

Disappearing Propeller Boat Co.
Port Carling, ON
Trade name: Dispro

Dispro. *See* Disappearing Propeller Boat
   Co.

*Easthope Bros.
Vancouver, BC
Trade name: East Hope

Etherington. *See* Atchison-Bower

Fleming Bros.
Halifax, NS
Trade name: (?)

*Foreman Motor & Machine Co., Ltd.
Toronto, ON
Trade name: Foreman

Fraser Machine and Motor Company
New Glasgow, NS
Trade name: Fraser

*Guarantee Motor Co.
Hamilton, ON
Trade name: Guarantee

Hamilton Motor Co.
Hamilton, ON
Trade name: Hamilton

Hawboldt Gas Engines
Chester, NS
Trade name: Hawboldt

Imperial. *See* Bruce Stewart & Co., Ltd.

J.W. Cumming Mfg. Co.
New Glasgow, NS
Trade name: (?)

Lloyd's Manufacturing & Foundry Co.,
    Ltd.
Kentville, NS
Trade name: Minas

Lunenburg Foundry Ltd.
Lunenburg, NS
Trade name: Atlantic

McKeough & Trotter Ltd.
Chatham, ON
Trade name: McKeough & Trotter

Midland Engine Works
Ontario
Trade name: Hanley Marine

Minas. *See* Lloyd's Manufacturing &
    Foundry Co., Ltd.

N.S. Steel & Coal Ltd.
New Glasgow, NS
Trade name: (?)

*Pacific Standard Motor Works
Vancouver, BC
Trade name: Pacific Standard

Peter Payette & Co.
Penetanguishene, ON
Trade name: Payette (?)

Pictou Foundry & Machine Co.
Pictou, NS
Trade name: (?)

St. Lawrence Engine Co.
Brockville, ON
Trade name: St. Lawrence

Schofield-Holden
Toronto, ON
Trade name: Schofield-Holden

Trask & Company
Halifax, NS
Trade name: Trask

Vivian Gas Engine
Victoria, BC
Trade name: Vivian

Appendix B

Today, when motor vehicles of all descriptions are taken for granted, it may be hard to imagine the basic questions and problems that faced a man 80 years ago, when confronted by the first gas engine he had ever owned. The following humorous tale will serve to remind us. It was published in Powerboat News *on May 13, 1905.*

FISHERMAN JOE AND HIS GASOLENE KICKER

by E.J. Williams

Fisherman Joe was a typical salt in bearing, speech, and habits. He lived in the town of Halibut, and managed his meagre existence by the sale of sea food at the adjoining town. Almost every day Joe could be seen rowing his skiff into the harbor, after a tiresome day's work. Rowing, especially on warm days, was the only part of the trip he was at all averse to, and whenever a favorable zephyr sprang up he set the sail. One day after Joe's arrival at his domain, a small hut near the shore, he observed a stranger ambling along, who presently stopped at the skiff and eyed her from stem to stern. As the stranger appeared in no hurry to depart, Joe swaggered down to look him over.

"How dy?" remarked Joe as a matter of salutation.

"Do you own this boat?" asked the stranger.

"Wall, I reckin."

"Use her every day?"

"Purty nigh every day."

"Row her, I suppose?"

"Wall, I ginerly hev tu, when there's no wind."

"Why don't you put a gasolene engine into her, and then you wouldn't have to row or wait for wind?"

"I suppose you have seen them?" remarked the stranger.

"Wall, yes, onc't or twic't, up yonder," pointing to the next town, "but I don't know nothing about makin' 'em go."

"Oh! that's dead easy. Push on the switch, turn over the wheel, and away you go until you want to stop, and all you have to do is sit in the stern and steer her."

"By gum! that must be all right," suggested Joe.

"You bet it is, just keep the tank full of gasolene, and you are always ready to go."

"Wall, say, neow where can I find erbout them things?"

"Why, I am in that business. My name is Smith, Bill Smith. I represent the Rapid Fire Engine Company. If you say so, I will send one here, and I'll bet when you see it you will want one. I will be here next week, and will try and have the machine here then," and shaking hands with Joe, Bill Smith departed.

Joe walked slowly home and imparted the scheme to his partner Silas. At first Silas said he didn't like "them new fangle ideas," but Joe said he had not bought it, and if he did not want it, didn't have to take it.

Wednesday morning of the next week arrived and Bill Smith had not yet put in an appearance. By this time Silas was as enthusiastic over his arrival as Joe. About noon a truck drove up with a heavy box, which was left at the door, and a few minutes after, along came Smith. After a short rest, Smith asked for a hatchet, and began to remove pieces from the box. Joe and Silas crowded for the first peep. After removing the box, Smith tore off the paper showing the "Rapid Fire," decorated in red enamel.

"Ain't she a beauty?" remarked Smith.

"Yew bet!" assented Joe, and Silas grunted his approval.

"Heow much be thet wurth?" queried Joe.

"One hundred dollars, put in the boat, complete. This is two and a half horse-power and will push your boat five miles an hour."

The latter part of the remark was lost, for Joe gasped for breath, and Silas grasped a twig for support, but lost his balance and fell backward over a stone.

"Be yew yarnin, or talkin strate?" ventured Joe.

"Well! I might make it ninety dollars and put the engine in the boat myself, as this is the first machine in town. Throw some canvas over it and I will return in an hour; and you can think it over in the meantime."

Fifteen minutes after Smith had departed Joe and Silas came to the conclusion that they wanted the machine, so when Smith returned, arrangements were made to have the boat in working order by the following Saturday afternoon. Smith started in the next morning, and by Saturday noon everything was in readiness for the trial. The tank was filled from the grocery store, and the boat pushed into the water. With Silas in the bow, and Joe at the tiller, Smith rocked the fly wheel back and forth once or twice, and off she started out on the bay, passing rowboats as if they were at a standstill. During the return trip Smith explained to Joe all the points necessary, and had him start and stop the engine a dozen times. After the boat was tied up Smith bade Joe and Silas good-bye, leaving his card with Joe, saying that in case anything occurred to telegraph, and he would come right down.

The next morning Joe and his partner went down for a spin. They jumped in and Silas cast off the bow line. Joe uncovered the engine, and following Smith's directions, the boat started. A smile of happiness crept over their faces, until about five hundred feet had been covered, when, "ker chung," and the engine stopped. Silas had let go the tiller, and Joe had backed toward the bow at such an unwarranted action.

"What the deuce ails it, Joe."

"Garl darned ef I know," retorted Joe, and gaining courage, he took hold of the fly wheel and turned it again. "Chug" — "chug" — "ker-chung." And with all the turning after that, not a "chug" would she give again. A very crestfallen pair they presented as they rowed the boat back to the mooring buoy.

Joe went to the telegraph office and dispatched the following telegram to Smith:

"Bill Smith, Engineville.

Rapid fire busted twice, come quick.

Fisherman Joe."

Smith arrived on the next train expecting to see the wreck of the boat, and the owners maimed and badly hurt, but was met by Joe and escorted to the boat, explaining all that occurred, especially the two busts.

"Now, Joe," remarked Smith, "show me what you did after 'she busted,' as you call it." Joe proceeded through the maneuvers of starting, but she did not go. "Now, Joe," again remarked Smith, "when she flunked with you, we call it back-firing, and is generally caused by insufficient gasolene supply. Put your finger under the vaporizer and lift that poppet and see if there is any gasolene. That's right. No gasolene," and turning around toward the bow, he exclaimed, "turn on that tank stop cock. You can't run without fuel. I told you to turn it on the first thing when you entered the boat. Now lift the poppet. Gasolene now? Good." And casting off from the mooring off she went at the first turn of the wheel. Smith took the next train home.

A storm came on the next day, and no opportunity was offered to use the boat for three or four days. As soon as it cleared Silas and Joe bailed her out, and decided to take a spin before putting her into use the next day. Joe was very careful this time to follow the directions, but go the engine would not. They tried it the next day, but with the same result, and as a last resort telegraphed for Bill Smith. Bill arrived promptly, and had Joe go over the maneuvers of starting again.

"Gasolene, Joe?"

"Yes."

"Spark?"

"Guess so."

"Don't you know," ejaculated Smith, with disgust, and disconnecting the wire from the insulated plug, he touched a cylinder head bolt.

"No spark, Joe," and taking a piece of sand paper from his pocket he ran it through the grooves of the knife switch, and rubbed the knife until it was bright. Closing the switch he touched the engine and a flame shot from the wire. He then screwed the wire in place, grasped the fly wheel and off she started.

Smith departed that noon in a thoughtful mood, after having told them to pour a half pint of oil in the crank case once a week, through a plug hole. Owing to bad weather the skiff was not used again until the next Saturday. Joe had by this time begun to think he had mastered all the kinks of a gasolene engine, and so, acting upon Silas' suggestion that, as the week was up, the crank case required oil, Joe began by taking out the plug in the side of the case and placing it in his pocket for safe keeping. The last bit of oil having disappeared from the can, preparations were made to start. He turned the fly wheel once, twice, three times. No, she would not start. He looked at the tank. Yes, plenty of gasolene. Tank valve turned on, vaporizer valve turned on, needle valve set at number 4, spark good. After Joe had churned for an hour, Silas said he thought he might start it. But no, the thing refused to budge. Must they send for Smith again? After another three hours' churning the fly wheel, Smith was sent for. He arrived with a cross look, and it was apparent he was in a bad frame of mind.

He watched Joe turn the engine over, try the gasolene, and the spark. Then he took the fly wheel himself.

"There's a leak in the crank case. I hear it," he growled, and leaning over to one side of the engine he fairly yelled, "where is that oil plug out of the case? It's gone." Joe fumbled in his pockets, and finally handed it over as meek as a schoolboy, and Smith lectured him on his loss of memory, while he replaced the plug. Smith did not wait for Joe to start, but jumped ashore and told them to go ahead, that she was all right. As he turned the corner he glanced back and saw them sailing down the bay with the engine running smoothly. As to Smith's meditations it would not be fair to speak.

The writer left the town of Halibut the next morning, but from recent communications from that town, Bill Smith still finds it necessary to run over every few days to start Joe's engine through some absentminded antic of the owner.

Appendix C

Although considered America's earliest automobile magazine, The Horseless Age *also addressed itself to the broader subjects implied by its title. The horseless age envisioned by the magazine included self-propulsion on water as well as land. In November 1900, the journal published the following article and plans of a gas engine for stationary or marine use. The engine represents the state-of-the-art at the turn of the century.*

ONE-HORSE-POWER GAS ENGINE

by Geo. H. Johnston

The construction of small engines for scientific instruction, or practical purposes, has been a source of much interest and study, and is a favorite subject with the amateur and those in need of light power.

For the benefit of those who wish to construct a small gas or gasoline engine, the following design and drawings are submitted.

The designer has made it a point to simplify the method of construction, and reduce the cost to a minimum, in order to enable the average amateur to construct it.

The various parts of the mechanism have been designed from a theoretical stand-point, and are proportioned for actual work, and the dimensions are the result of calculation and tests. It has been the aim to make the engine as compact as possible, and to illustrate it by detail drawings in a practical manner.

This engine is of the two-cycle type, receiving an impulse at each revolution of the crank, and igniting at constant volume, with previous compression. Gas and air are drawn into the crank case by the upstroke of the piston. On the downstroke the mixture is slightly compressed, the inlet port is uncovered, and the mixture is forced from the crank case into the cylinder. On the return stroke this port is covered by the piston, the charge is compressed to one-third its original volume, and is exploded near the end of the stroke, by a spark from the electric igniter, operated by the piston. As the piston moves downward, due to the expanding gases, the exhaust port is uncovered, thereby allowing the burnt gases to escape.

It will be observed that the exhaust port is uncovered just before the inlet port, which is at the end of the stroke, and that both ports are open for a short period of time. The fresh charge is turned upward into the cylinder, by means of the deflector on the piston, opposite the inlet port. In this way the greater portion of the burnt gases are displaced. It will readily be seen from the above explanation that the engine can be run in either direction.

The engine will run at 550 revolutions per minute. With close attention to drawings, and good workmanship, this engine will develop one brake horse-power.

The drawings illustrate both marine and stationary engines. It will be noted that the two engines are similar, excepting the base, flywheel and governing mechanism, which are additional on the stationary engine.

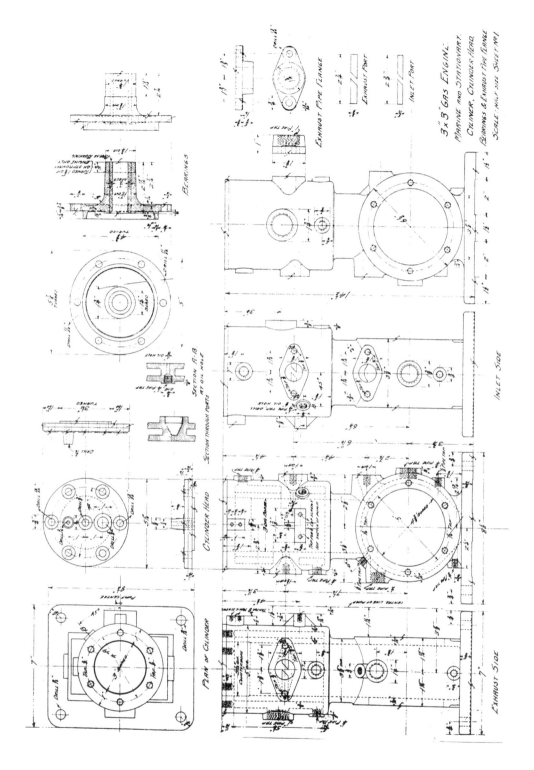

3"× 3" GAS ENGINE:
MARINE AND STATIONARY.
CYLINDER, CYLINDER HEAD.
BEARINGS & EXHAUST PIPE FLANGE
SCALE : HALF SIZE   SHEET N° 1

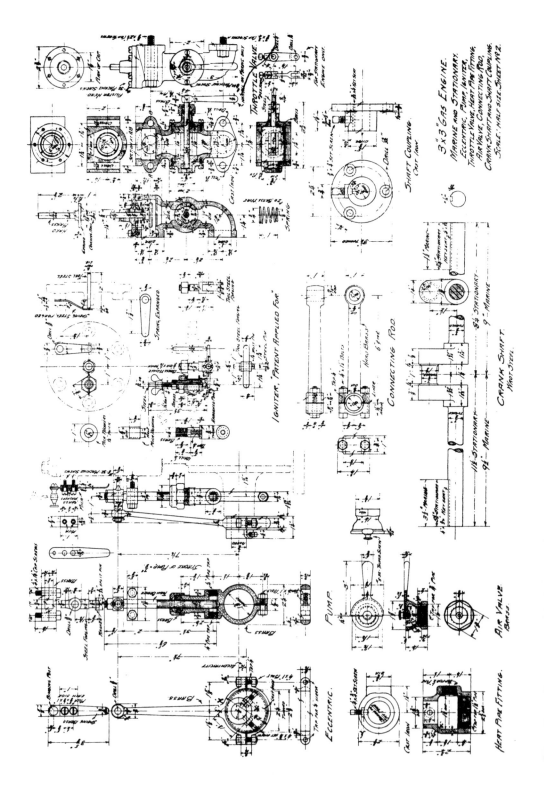

3"x3" GAS ENGINE.
MARINE AND STATIONARY.
ECCENTRIC, PUMP, IGNITER,
THROTTLE VALVE, HEAT PIPE FITTING,
AIR VALVE, CONNECTING ROD,
CRANK SHAFT AND SHAFT COUPLING.
SCALE: HALF SIZE. SHEET N°2.

THROTTLE VALVE.

SHAFT COUPLING.
CAST IRON.

CONNECTING ROD.

CRANK SHAFT.
WROT STEEL.

IGNITER. "PATENT APPLIED FOR"

PUMP.

AIR VALVE.
BRASS.

ECCENTRIC.

HEAT PIPE FITTING.

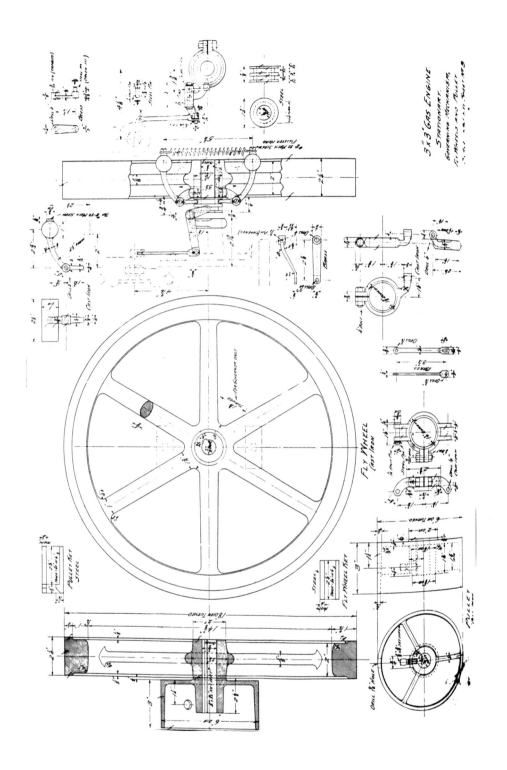

3"x3" GAS ENGINE
STATIONARY,
GOVERNING MECHANISM,
FLY WHEEL AND PULLEY

FLY WHEEL
(CAST IRON)

PULLEY KEY
STEEL

FLY WHEEL KEY
STEEL

PULLEY

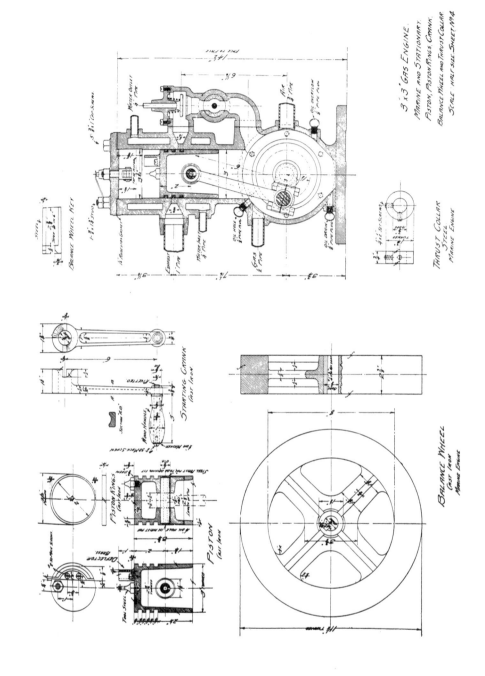

3"x 3" GAS ENGINE.
MARINE AND STATIONARY.
PISTON, PISTON RINGS, CRANK.
BALANCE WHEEL AND THRUST COLLAR.
SCALE HALF SIZE. SHEET N°4.

BALANCE WHEEL KEY

THRUST COLLAR
STEEL
MARINE ENGINE

STARTING CRANK
CAST IRON

PISTON RINGS
CAST IRON

PISTON
CAST IRON

DEFLECTOR
BRASS

BALANCE WHEEL
CAST IRON
MARINE ENGINE

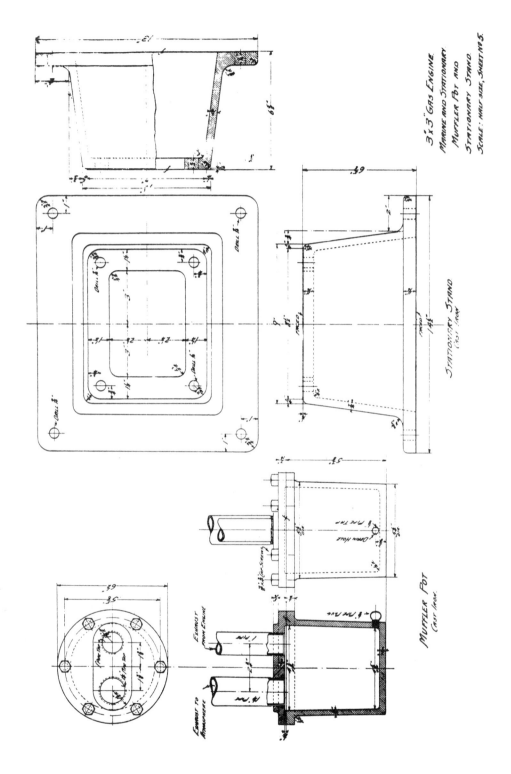

3"x 3" GAS ENGINE.
MARINE AND STATIONARY
MUFFLER POT AND
STATIONARY STAND
SCALE: HALF SIZE, SHEET N° 5.

STATIONARY STAND
CAST IRON

MUFFLER POT
CAST IRON

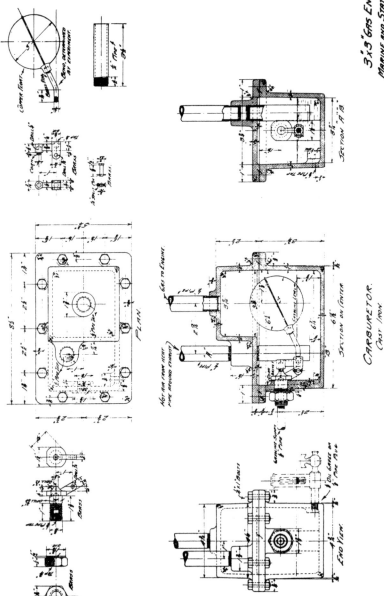

3"x3" Gas Engine.
Marine and Stationary.
Carburetor and
Fittings; Complete
Scale : Half Size. Sheet No 6.

Carburetor.
Cast Iron

Plan

Section on Enter

Section A B

End View

Sheet No. 1.

The cylinder, water jacket and crank case consist of one casting, and should be carefully machined to dimensions shown, special care being taken to have bore of cylinder true and smooth.

The bearing plates, which fit on each side of the crank case, are supplied with brass bushings. These bushings should be neatly fitted and kept from turning by a small dowel-pin. The bearings do not require oil cups, as the oil splashed about by the connecting-rod, in the crank case, furnishes the required lubrication; however, when the engine is new, they should be liberally supplied with oil for a time. A paper gasket should be placed between bearings and crank case in fitting up.

The upper edge of the cylinder is intended to be finished. The cylinder head is provided with the necessary bosses to accommodate the igniter and should be faced, and fitted to the cylinder, and the cylinder should be turned with the head in place, thereby giving it a neat and finished appearance. Use a $\frac{1}{16}$-inch-thick gasket between cylinder and head.

The exhaust pipe flange is of the usual pattern, and should be fitted to the cylinder with an asbestos gasket between faces.

The detail of the exhaust and inlet ports shows the manner of bridging to allow free passage of piston rings.

Sheet No. 2.

The eccentric is provided with set screws, and so arranged on the shaft that the cross-head on the pump plunger will make contact with the flat spring outside, before the piston breaks the contact inside, and contact is broken outside, before being made again inside. The tip of the spring should be slightly bent to facilitate this adjustment. This spring should be carefully insulated with mica in fitting same to cylinder, and provided with a binding post, as shown.

The circulating pump, which forces water into the jacket, for cooling the cylinder, is operated by the eccentric, and should be fitted with a check valve on each side. The plunger should be packed with candle-wicking, or other suitable packing.

The igniter is operated by the piston, which causes the double-armed lever to break contact with the insulated electrode fitted in the cylinder-head, and projecting into cylinder. This electrode should be insulated with mica bushing and washers, as shown. The adjusting screw, which allows the timing of the spark (which should occur just before the end of the up stroke), should have a hardened point. Drawings show the spark coil and battery connections. Use closed circuit batteries with a 10-inch spark coil.

The throttle valve is in the conduit, which connects the crank case with the cylinder. The speed of the marine engine is controlled by a handle attached to the stem of the throttle valve.

On the stationary engine this handle is replaced by a small lever, which connects with the governing mechanism.

The poppet valve, above the throttle valve, prevents back firing into the crank case. The valve should be ground on its seat, and kept in place by a light brass spring.

The conduit or gas passage should be fitted to the cylinder and the crank case with paper gaskets between faces.

The heat pipe fitting fits over the exhaust pipe, and is secured by a set screw. The heat pipe, if possible, should be from 15 inches to 24 inches long.

The air valve permits of variable opening to suit the condition of the atmosphere, and is provided with a thumb screw to securely hold it in position after adjustment.

The connecting rod is intended to be made of hard brass, and requires, therefore, no bushings. The bolts in the crank end of the rod should be screwed in, the nuts thereby acting as checks and locking the bolts in place.

The crank shaft should be finished all over, and be key seated, as shown.

Shaft couplings are shown for the marine engine, for connecting the crank to the propeller shaft.

Sheet No. 3

The stationary engine is provided with two flywheels. To one of these the governor is fitted. The governor is of the ordinary flywheel type, and connects with the lever on the stem of the throttle valve. The drawing fully illustrates the action of the governor. Care should be exercised in adjusting the throttle and governing mechanism. The strength of the coil spring should be carefully determined, by experiment, to suit the existing conditions.

The driving pulley should be keyed and fitted on the end of the crank shaft after the flywheel is in place. The head of the key should extend beyond the hub of the driving pulley, and the key be secured by a set screw.

Sheet No. 4.

The deflector and the igniter point are to be fitted in position on the piston as shown. The connecting rod should be fitted in the piston, and the wrist pin made a driving fit, and provided with the necessary oil holes.

The piston rings should be turned eccentric, and slipped into place with the aid of thin strips of sheet iron or other available metal. The starting crank is provided with notches in the hub, which engage with a key on the end of the shaft. The key projects through the hub of the driving pulley.

The balance wheel on the marine engine should be firmly keyed to the shaft, and the rim finished bright.

The thrust collar is secured to the shaft with set screws, and transmits the thrust to the ball thrust bearing.

The section taken through the center of the engine fully illustrates the engine and all internal parts.

Sheet No. 5.

The stand on which the stationary engine is mounted should be properly secured to a suitable foundation.

The muffler reduces the noise of the exhaust, and allows the gases to expand somewhat before reaching the atmosphere.

Sheet No. 6.

The carburetor supplies vapor from gasoline for the marine engine, and for the stationary engine when it is located in an isolated place, where illuminating gas is not available.

The action of the carburetor is apparent from the drawing. Gasoline flows by gravity from the supply tank, situated in the bow of the boat in the case of marine engines, or at some convenient point outside of the building in the case of the stationary engine.

The action of the copper float tends to keep a uniform level of gasoline at all times. All joints and connections should be tight to insure good results.

An oil gauge to indicate the amount of gasoline in the carburetor can be fitted, at the option of the builder. If gauge is not used carburetor should be provided with drain hole, as shown.

Appendix D

Books on gas engine operation and care proliferated as did the engines themselves. The following, from Marine Gas Engines, Their Construction and Management, *was written by Carl H. Clark and published by D. Van Nostrand Company in New York in 1914. It is noteworthy for its straightforward conveyance of information and should be as useful to an enthusiast or collector as it was to an engine owner of its own era.*

CHAPTER XII. Operation and Care of Engines

Starting. — Before attempting to start an engine one should acquaint himself with the operation of all parts of the engine, and the water, oiling, and ignition systems.

If the engine has been standing some time, the ignition system should be tested to make sure that it is in good order. The make-and-break system may be tested by first closing the switch in the circuit, then turning the flywheel until, by observing the action of the mechanism, the points inside the cylinder are known to be in contact, closing the circuit. The wire is then removed from the insulated terminal of the igniter and brushed across it; if the circuit is complete, a brilliant spark will result. If no spark is obtained, all connections must be examined and, if necessary, the sparking points removed from the cylinder and cleaned. After a spark has been obtained through one cylinder the other cylinders should be tested.

The jump-spark system is tested by removing the plug from the cylinder and resting it upon any of the bright metal parts of the engine with the secondary wire still in contact. With the primary circuit closed the engine is then turned until contact is made by the timer, which should be indicated by the buzz of the vibrator. The sparks should pass across the points of the plug at the same time. If no buzz of the vibrator occurs, it shows a defect in the primary circuit. If the vibrator buzzes, but no spark passes the sparking points, it shows a fault in the secondary circuit. This may often be remedied by removing any deposit of carbon or oil which may have collected on the points of the plug, or by fitting another plug. After a rain, or in damp weather, the most common sources of ignition troubles are weak batteries or a "ground." The former can be found by testing the batteries with an ammeter; the latter is caused by the current, especially the secondary, jumping across from the wire to some adjacent pipe or other part. It may be found by carefully examining the wires, or, in the case of the secondary current, the sparks may be seen to pass where the current jumps. The wiring of each cylinder should be tested in turn. At the same time that the ignition is tested, the point of ignition should be noted in reference to some point on the flywheel, and its variation with the different positions of the timer handle noted. The gear or timer should be so set that the spark occurs when the piston is at the top of its stroke.

Oil- and grease-cups should now be filled and a small amount fed from each.

The fuel supply should now be turned on, both at the tank and at the carburetor. The needle valve on the carburetor or vaporizer should be opened slightly and the carburetor primed somewhat to make sure of a good flow of fuel. The engine may now be turned over by hand in the direction in which it is to run, using the crank, lever, or flywheel rim, as the case may be. After a few trials the engine should explode a charge and turn a few turns and possibly continue to run. If it does continue to run, the fuel and air supplies should be adjusted gradually until the engine turns at its highest speed. Oil-cups should then be opened to allow the oil to feed.

If, in the case of a four-cycle engine, it does not start at once, all that is necessary is to turn it under varying conditions of fuel and air supplies, after making sure that the ignition system is operating and that fuel is flowing to the carburetor. In the case of a two-cycle engine the fuel should be shut off and the compression cock opened. It is probable that several charges of gasoline have been taken into the base and not exploded, making the mixture far too rich and "flooding" the engine, as it is termed. The flywheel is now turned several times and the mixture diluted and partially expelled. An explosion will finally take place and the engine run until the supply in the base has been used up.

Another trial can now be made, with a reduced fuel supply, and continued under varying conditions until the engine starts, always making sure that the ignition occurs properly, and taking care not to flood the base.

Starting is often made easier by priming the engine, that is, by inserting a small amount of gasoline directly into the cylinder through the compression cock, or through a special priming cock which on some engines is provided for that purpose.

When turning the engine over by hand, care must be taken to have the sparking gear so set that the spark cannot occur until the piston has reached the top of its stroke. If this is not done and ignition takes place before the piston has reached the top of its stroke, it will be driven violently downward in the wrong direction, giving a "back kick," which is liable to cause damage.

After the engine has run a short time it will often gradually slow down and finally stop, with a muffled explosion in the base. This is a sign of a weak mixture, and the fuel supply should be slightly increased. On the other hand, if the engine labors, slows down, and finally stops, with black smoke issuing from the exhaust, it shows a too rich mixture and the fuel supply should be cut down.

The best fuel mixture can be found only by experiment, and will even then vary somewhat, according to atmospheric conditions. The proper regulation of the air supply to the fuel supply has a marked influence upon the fuel economy.

Two-cycle engines of the two-port type are easily and readily started as follows: The spark is advanced to a point somewhat below the usual running point so that the ignition will take place on the up stroke when the engine is turned in the opposite direction to which it runs. The flywheel is then turned until the piston is at the bottom of its stroke, and is then rocked backwards and forwards a few times; the piston thus acts as a pump, drawing a few small charges into the base and charging the cylinder. The flywheel is then turned quickly in the reverse direction to which it runs, bringing the piston up against the charge, which finally ignites and forces the piston down again, but in the right direction. The flywheel is then released and the engine starts. The spark is then restored to the running position.

This method cannot be used on a three-port engine, which must be started by turning it over the center.

Some engines will start with the compression cocks fully or partly open, which lessens the labor; others must be pulled over against the full compression.

In starting the engine with a crank or lever it should be held loosely in the hand, so as to be quickly released in the event of a back kick. Frequent accidents happen from the disregard of this precaution.

Oiling. — When the engine is well started the fuel and air supplies should be regulated until the engine is running on the least possible amount of fuel. The oil supply should be regulated to a point just below that at which smoke would issue from the muffler. Blue smoke coming from the muffler usually shows that too much oil is being fed, which instead of being of use in the engine is burned or carried away by the exhaust, and wasted.

The cylinder oil-cups should be adjusted to feed from three to six drops per minute, according to the size of the engine. Where splash lubrication is used, oil must be fed into the crank case or base at intervals. Exterior parts, such as thrust bearing, igniter gear, and pump journals, are, of course, oiled from an oil can when necessary. In running a new engine, oil should be used rather freely at first while the bearings are wearing down into place. The cylinder surface may be greatly improved by feeding in some powdered graphite mixed in oil, which fills up the pores and helps to form a sort of scale on the bore of the cylinder. While too much oil should not be fed, as it is not only wasted, but makes the engine dirty, a sufficient lubrication should be made certain at all times, as much damage may be done in a short time if bearings are allowed to go dry.

Under some circumstances good results can be obtained by mixing the lubricating oil with the gasoline in the tank and feeding both together. No difficulty is experienced with the vaporization and the lubrication is simplified. Although the relative amounts will vary considerably, a fair proportion seems to be about one pint of oil for every five gallons of gasoline.

Spark Advance. — While running the engine it will soon be noted that the time of ignition has a great effect on the speed. It will be found that the engine runs best when the ignition takes place just before the piston reaches the top of its stroke. This is due to the fact that the burning of the charge is not instantaneous, but requires an appreciable time. If the charge is fired at the moment when the piston is at the top of the stroke, the time taken by the charge to thoroughly ignite allows the piston to descend through a part of the down stroke, so that some of the effect of the impulse is lost. If the spark is so timed in advance that the charge is completely ignited at the time when the piston is just ready to descend, the full effect of the impulse is received and absorbed.

This advance of the spark is called the "spark advance" or "lead." It will of course vary somewhat according to the speed. The speed of the engine can be varied by shifting the point of ignition, and this is advocated by many. Starting with the spark occurring as the piston is at the top, it will be found that up to a certain point the speed will increase as the spark is advanced. Beyond this point the engine will pound and act irregularly. If the spark is retarded until after the piston has begun to descend, the speed will decrease. The speed of the engine may thus be regulated by changing the spark advance, but this practice is not to be recommended, as nearly the same amount of fuel is passed through the engine per stroke at all speeds. At low speeds the charge ignites so slowly that all the heat generated cannot be absorbed, but is passed along into the exhaust pipe and muffler, heating them beyond their usual temperature. The speed of the engine should be regulated by the throttle which is usually provided for that purpose; in this way the amount of the mixture is cut down in proportion with the speed. The speed should be regulated by the throttle and then the spark advanced to the best point by trial. In this way the greatest economy in the use of fuel may be obtained. Extremely slow speed must, however, be obtained by retarding the spark, in connection with the throttle.

Care of the Engine. — The degree of care which the engine receives, not only when running, but when laid up as well, has a great effect upon its life and also its satisfactory operation. Many engines which are well taken care of while in operation are allowed to suffer from exposure during the time when they are not in use. If the engine is in a cabin boat it is very easily kept in good shape, but if in an open boat, constant care is required to prevent it from being damaged by rust. A cover should be made from waterproof canvas, which will fit snugly over the engine and shed all rain. A water-tight pan under the flywheel will prevent the bilge water from rising around it and causing it to rust. Before leaving the engine for a few days all bright parts of iron or steel should be smeared lightly with grease, which may be readily removed with cotton waste. This precaution will save a large amount of scouring and polishing later.

One should become thoroughly familiar with the construction of his engine as soon as possible. It is not meant by this that the engine should be pulled down just to see how it is constructed, but quite the contrary. As long as the engine is running, particular pains should be taken not to disturb it. The construction should, however, be studied so that in case of necessity it could be taken down. Much expense can often be saved by this knowledge, as there are many small repairs which can easily be made by the amateur owner.

Before starting on a run all nuts and bolts should be examined, and any which may be loose should be tightened.

Engine Troubles. — The presence of trouble in the engine is usually indicated by a peculiar hammering noise, known as a "knock." It may be caused by excessive friction on some part and the oiling system should be at once examined and perhaps an additional amount fed for a few moments. A similar knock may be caused by the failure of the water-circulating pump, which may be told by the unusual amount of heat radiated from the cylinders. The lack of cooling water causes the cylinders to become much too hot for use, increasing the friction and eventually causing damage to cylinders and pistons. If the knock cannot be found in this way, the engine should at once be stopped, as damage may be caused. The knock is probably caused by some part which has become loosened, and all parts should be thoroughly examined. A loose flywheel is a common cause of knocking. Where the flywheel is fastened with a key, the keyway may become worn so as to leave a small space between it and the sides of the key, allowing the flywheel to "play" slightly around the shaft. If this is the case the key should be withdrawn and a slightly wider one fitted, or a thin "shim" of steel may be carefully fitted into the keyway and the key driven in alongside of it.

A bearing which has become overheated and ground out will cause a knock; this is a more difficult cause to remedy, and is likely to require the services of a machinist to reset the bearing.

Sometimes the engine will run with apparently no trouble, and yet will show less than the usual power. This may be due to loss of compression, or in other words, a leakage from the compression space. This may be due to a loose plug or screw at some point, or in the case where the cylinder head is fastened on with studs, it may mean that the gasket under the head has become broken at some point and it may be remedied by fitting a new gasket. If the cylinder has been flooded too freely with oil, the excess may carbonize and collect around the piston rings, cementing them to the piston and allowing the gas to escape by the piston. This may often be remedied by flushing the cylinder with kerosene oil. It may often be necessary to remove the cylinder and separate the rings from the piston. In removing the rings from the piston great care is necessary as they are of cast iron and very brittle. Before attempting to remove them they should be well washed with kerosene, to loosen them as far as possible. In removing a ring, one corner should first be raised with a screw-driver or other tool, and a narrow strip of tin placed across the groove under it to keep it from springing back; this is followed up all around the ring, tapping it lightly and adding more strips until the ring is entirely supported clear of the groove. It may then be slid from the piston. Rings and grooves should be thoroughly cleaned with the help of kerosene. In replacing the rings the reverse operation is followed.

In the four-cycle engine it is necessary to "grind in" the valves at intervals when the seats and surfaces become pitted or worn. When the valves are arranged, the inlet valve may be removed to allow access to the exhaust valve. The springs are removed from the stems, the valve is raised and the bevelled edge smeared with a paste of oil and emery. The grinding is done by rotating the valve in its seat by means of a screw-driver or brace; the process is continued until the surfaces are left smooth and polished with no sign of corrosion. This paste is then removed and the finish put on with a mixture of water and pumice. During the grinding the entrance to the cylinder should be carefully stopped with a wad of waste to exclude the emery from the cylinder, where it would do great damage. If the inlet valve is removable it may be ground while held in the hand.

In replacing the valves it may often be found difficult to compress the springs sufficiently to allow them to be replaced. If no other means is at hand the spring may be compressed in a vise and bound with a few turns of strong cord. It may then be slipped on the stem and the key and washer put on. The string may then be cut and the spring let out.

Care of Spark Coil and Ignition Outfit. — The entire ignition outfit is somewhat delicate and requires its share of attention. It should be examined at intervals to make sure that all binding screws are tight and all contacts good. The insulation should be examined and any places where it becomes worn should be taped. Two parallel wires should never be fastened by a single staple, as the insulation is likely to chafe through, causing a short circuit. Each wire should have its separate staples.

The most common place for a wire to break is at some place where it is bent back and forth, as where the wires are connected to the timer; a short coil at such points will greatly increase the life of the wire.

The entire system, including the plugs and sparking points, should be kept clean and free from oil. Oil on the outside of the plug will cause a short circuit, as does the collection of soot around the points.

The spark coil should be looked over and if necessary readjusted slightly. Many operators set the vibrator adjusting spring too tight, with the idea that the very rapid motion gives a stronger spark. This may seem true when tested in the atmosphere, but on a quick running engine it may give trouble by skipping. A more reliable spark is given with a moderately rapid vibration of the buzzer, with considerably less battery consumption. A satisfactory adjustment of the vibrator may be obtained as follows: The vibrator adjusting screw is drawn back until it is clear of the spring; the spring is then set so that the iron button is from $\frac{1}{16}$ to $\frac{1}{8}$ inch from the end of the core. The adjusting screw is then screwed in until it touches the spring lightly. The engine is then started and the screw turned in slightly until the engine runs steadily; the spring should be left as weak as possible and still have the engine run steadily. The spring must bear against the screw when the engine is not running, as otherwise no current will pass and the engine will not start. In testing the spark in the secondary circuit, the spark should not be drawn out to the limit, as this strains the coil and is almost sure to cause trouble if there is any weakness in the coil.

Batteries. — The batteries also should be examined occasionally. Weak batteries may be located by testing with an ammeter. When new, dry batteries of the usual size should test from 18 to 25 amperes, and as a rule they should not be used after they have fallen below about 8 amperes. A single weak battery will spoil the action of an entire set; even if no new ones are at hand the action will be improved by cutting out the weak one. In testing a battery the ammeter should be held across the terminals only long enough to get the reading, as if held there even for a short time the battery is quickly run down.

Batteries which have been run out may be temporarily revived by punching a hole in the top and pouring in some water.

Index

COMMUTATOR

Probably the simplest and most reliable commutator that has ever been built., We have used this commutator now for three years with very little change, just a few refinements, and commutator troubles are almost an unknown thing with the GRAY Motor Company.

It is elevated and enclosed so that any spark from it cannot set fire to the gasoline that may leak into the boat. It is provided with a water-proof, dust-proof oiler.

BRONZE GEAR PUMP

We have now in use over 16,000 motors equipped with bronze gear pumps. The very finest workmanship is employed in these pumps. Still some manufacturers have had trouble with gear pumps. It is because they use poor workmanship and poor material.

The gear pump must be made accurately and made right, then it is the best pump in the world.

We absolutely guarantee these pumps in every respect and claim that they give more water, a greater supply of water, than any other pump built.

While it is true that they are more expensive to build, they are quieter and neater, more reliable and better in every way.

GEAR COVER

The gears are carefully encased in tight coverings to prevent waste or rags being cut up in the gears and oil being thrown around the engine and boat.

CUT   GEARS

The gears used to drive the commutator-shaft are accurately and perfectly cut in automatic gear cutters in our own plant, so that the gears are necessarily quiet running and permit perfect turning.

TAPER CRANK SHAFT

The 1911 GRAY motor fly-wheel is attached to the crank shaft on a taper held securely by a lock nut. The taper is ground and the fit is perfect. The fly-wheel runs perfectly true and a perfect fitting key is also used.

LONG BEARINGS

Note the very long bearings used on GRAY motors. This is accurate and drawn to a scale and gives you a proper idea of the length of these bearings.

These bearings properly fit on a properly ground shaft using the highest grade bearing metal that money can buy, is another one of the reasons for the long life and high power of GRAY motors.

CONNECTING ROD BEARING

Are of the best babbitt, our own formula—better than ordinary bronze; costs us nearly twice as much, and will NOT CUT YOUR CRANK SHAFT as bronze will if you fail to lubricate it.

These bearings are interchangeable and, if through lack of lubrication one is cut or worn, you can replace it yourself in a few minutes.

CYLINDER

It takes a perfect cylinder to make a perfect gas engine.

GRAY cylinders are made of the best cylinder iron obtainable.

One of the reasons that GRAY motors hold their compression is because these cylinders are heated to a red heat after they are bored and reamed, then allowed to cool and then they are ground to a perfect round and finished surface. The heat process prevents the cylinder from distorting and changing its shape after it is subjected to the heat from exploding gas.

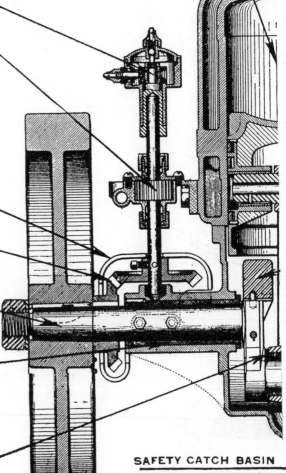

SAFETY CATCH BASIN

If you accumulate any foreign substance that might cause damage to your motor, this is a very important feature.

W9-CKH-729

FABLES
AESOP
NEVER
WROTE

but ROBERT KRAUS did

For
Parker

The collage illustrations are constructed from drawings, photographs, colored paper, stickers, a postage stamp, wax grapes, a stick of gum, a tea bag, a toothbrush, paintbrush bristles, cotton, glitter, artificial leaves, and faux pearls.

VIKING
Published by the Penguin Group
Penguin Books USA Inc., 375 Hudson Street, New York, New York 10014, U.S.A.
Penguin Books Ltd, 27 Wrights Lane, London W8 5TZ, England
Penguin Books Australia Ltd, Ringwood, Victoria, Australia
Penguin Books Canada Ltd, 10 Alcorn Avenue, Toronto, Ontario, Canada M4V 3B2
Penguin Books (N.Z.) Ltd, 182-190 Wairau Road, Auckland 10, New Zealand

Penguin Books Ltd, Registered Offices: Harmondsworth, Middlesex, England

First published in 1994 by Viking, a division of Penguin Books USA Inc.

1 3 5 7 9 10 8 6 4 2

Copyright © Robert Kraus, 1994
All rights reserved

LIBRARY OF CONGRESS CATALOGING-IN-PUBLICATION DATA
Kraus, Robert.
Fables Aesop never wrote / written and illustrated by Robert Kraus. p. cm.
Summary: Fifteen original fables, including "Sour Crêpes,"
"Fox in Chicken Feathers," and "The Dressy Wolf."
ISBN 0-670-85630-4
1. Fables. 2. Children's stories, American.
[1. Fables. 2. Short stories.] I. Title.
PZ7.2.K74Fab 1994 [E]—dc20 94-10936 CIP AC

Printed in China
Set in Frutiger

Fables

The ants worked hard all summer
storing food for the winter,
while the grasshopper played
his fiddle and sang,
"Hip, hop, I just can't stop
I'm a cool grasshopper
and I like to bop!"

A talent scout discovered him,
and now he's a rock star
on TV, while the ants are
still slaving away.

MORAL:
Practice,
practice,
practice.

The Kind Lion and the Brave Mouse

A lion caught
a mouse one day.
"Oh please don't eat me,
Mr. Lion," begged the mouse.
"And someday I will save your life."
"Fat chance," said the lion.
"But I have a kind heart
and I will spare your life."

Many years later, a hunter
was about to shoot the lion.
The mouse was passing by
and took the bullet himself.
"My brave bodyguard," cried the lion,
and shed a tear.
"Don't cry," said the mouse.
"I'm wearing a bulletproof vest."

MORAL:
A mouse could save your life.

The Fox, the Hedgehog, the Goose, and the Duck in the Manger

The fox, the hedgehog, the goose,
and the duck fell asleep
in a manger filled with hay.
The fox, the hedgehog, the goose,
and the duck would not move
to let the cattle eat the hay when
they came in, tired and hungry
from working in the field.

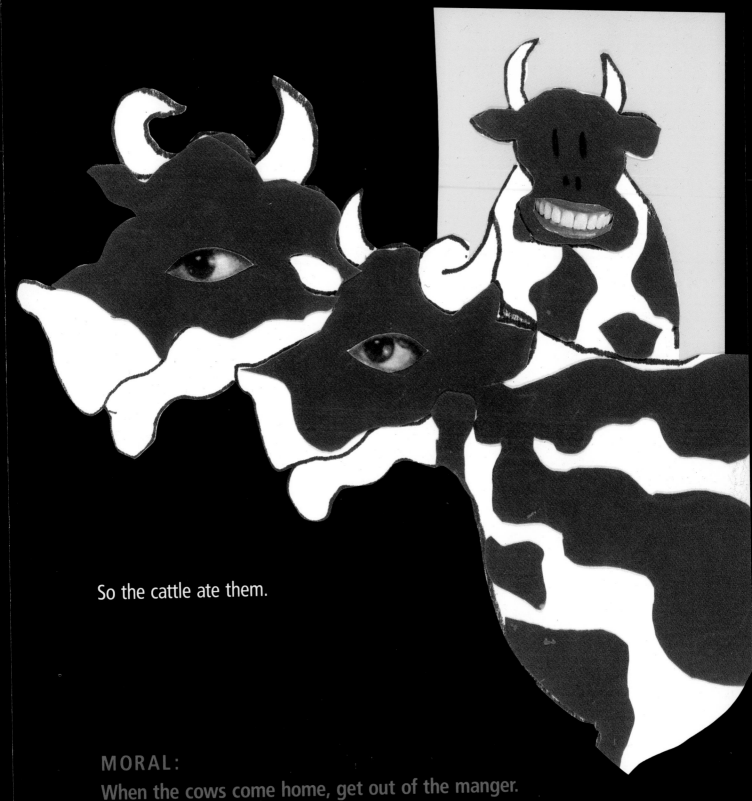

So the cattle ate them.

MORAL:
When the cows come home, get out of the manger.

The Dressy Wolf

There was once a wolf
who dressed in sheep's clothing
in hopes of fooling sheep.
But he fooled nobody.

So he dressed up in many costumes,
including a policeman's uniform.
He was arrested for impersonating
an officer and thrown in jail.

MORAL:
Dress up, but not as a policeman.

A farmer once had a goose
who laid golden eggs.
The farmer put all the golden
eggs in a basket to take to town to sell.
"I will be a rich man," he hummed.
He was robbed
by a highwayman who took all his golden eggs.

"Thank goodness I left my goose
at the farm," he sighed.
Hearing this, the robber hurried
to the farm and stole the goose as well.

MORAL:
Don't put all your golden eggs in one basket.
AND
Button your lip.

Two Masks

Two masks lived in the desert.

"What time is it?" asked Mask Number One.

"I don't know," said Mask Number Two. "My watch is bent."

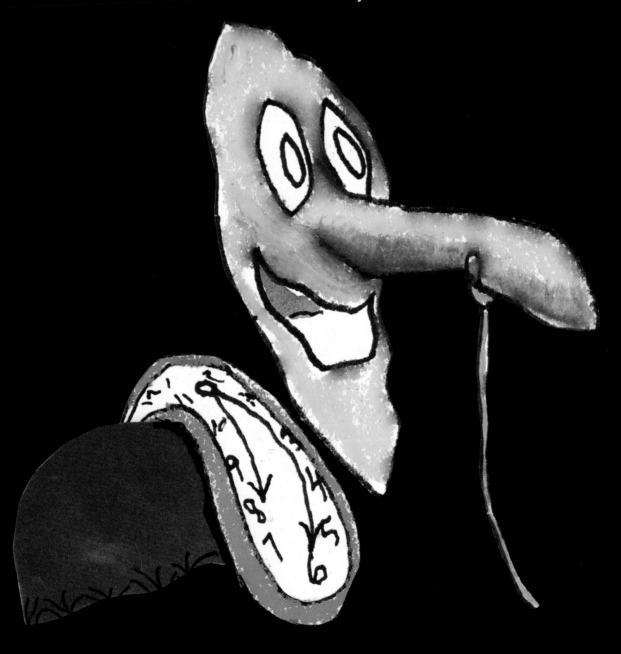

"So is your nose!" said Mask Number One.

"Well, if that isn't a case of the pot calling the kettle
black!" said Mask Number Two.

MORAL:
Masks don't have any brains.

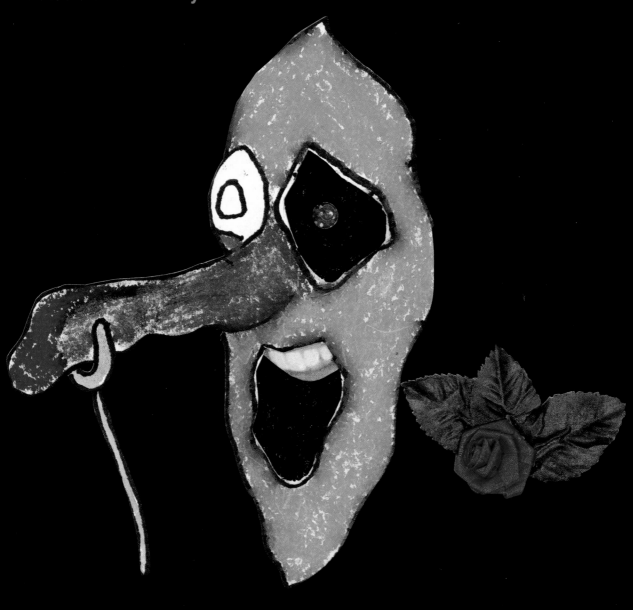

Lobster, Crab, and Shrimp Step Out

A lobster, a crab, and a shrimp
went to a fine restaurant for a meal.
"Do you serve fresh seafood?"
the lobster asked the maître d'.
"Now we do," said the maître d',
and introduced the three to the
chef, who dropped them into a pot
of boiling water.

MORAL:
Never inquire about the Catch of the Day.
It may be you.

The Fox in Chicken Feathers

A fox dressed in chicken feathers,
the better to sneak into the barnyard
and steal a chicken.

Another fox, thinking he was a large chicken,
stole him and had him for dinner that night.

MORAL:
Do not pretend to be what you are not.
AND
If you dress like a chicken, your goose is cooked.

The Dog and the Bone

A dog to whom the butcher
had tossed a bone
was hurrying home.
As he crossed a bridge,
he saw a bigger dog
with a bigger bone in the water.

He dropped his bone
and snapped up
the bigger bone.
He hurried home happily.

MORAL:
Go for it!

The North Wind, the Sun, and the Cyclone

The North Wind and the Sun
were arguing about
which one was stronger.
"I can make that traveler remove
his cloak," said the North Wind.
"I can make that traveler remove
his cloak," said the Sun.

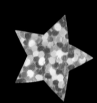

While they were arguing, a cyclone
removed the traveler, his cloak,
and the North Wind and the Sun.

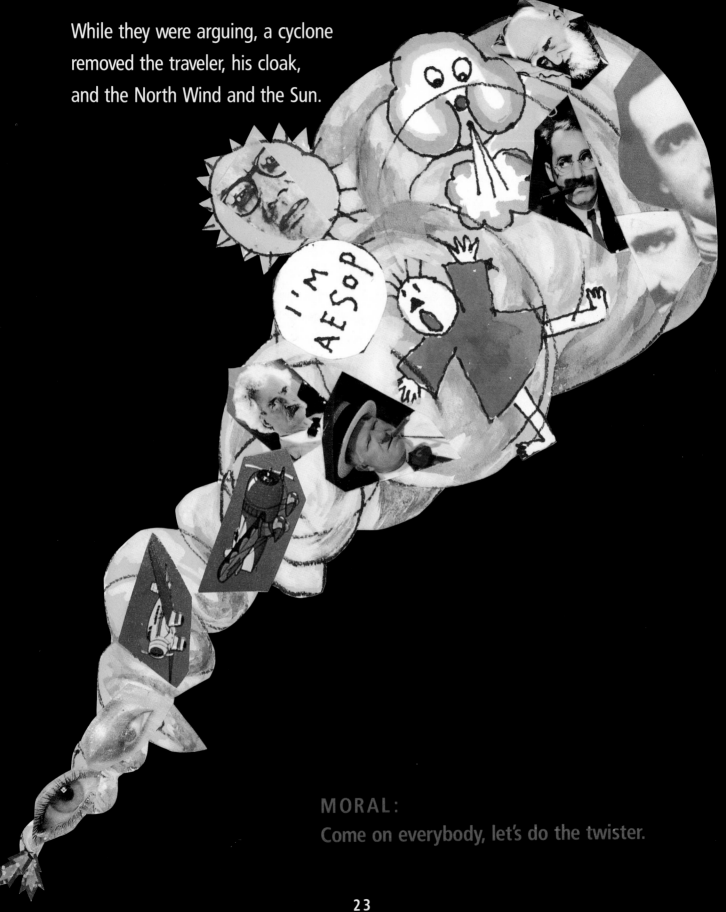

MORAL:
Come on everybody, let's do the twister.

Sour Crêpes

A baboon was making crêpes
for his lunch, when a
greedy fox came by.
"Mmmmm, those crêpes smell
good," said the fox.
"Can I have a taste?"
"No way!" said the baboon,
and climbed a tree to eat
his crêpes in peace.

The greedy fox jumped and jumped,
but could not jump high enough
to grab a crêpe.
"Probably those crêpes are sour,"
he said, as he plucked a bunch
of sweet grapes.

MORAL:
Red foxes can't jump.

"Boy!"
"Boy! Come here"

The Wolf Who Cried "Boy"

A wolf was hungry
and he wanted to make
a Poor Boy sandwich.
He had the bread, but not the boy.
"Boy!" he called. But nobody answered.
Little Boy Blue was fast asleep.
"Boy! Come here," he called again.
No answer.
"Boy! Boy! Boy!"
Still no answer.

So he picked up the phone
and he ordered a pizza,

and ate
the delivery boy.

MORAL:
Where there's a wile there's a way.

The City Moose and the Country Moose

A city moose was upset
by all the crime and shooting
in the big city.

He was happy to be invited
to visit his cousin in the country.
"Now for a little peace and quiet," he thought.

No sooner had he arrived than hunters
began shooting at him.
Bang! Bang! Bang!

"It's the season," said the country moose.

"I'm going home," said the city moose.

"I'm safer there."

MORAL:
Life is not easy if you're a moose.

Morals Without Fables

Don't judge a duck by its waddle.

A fast rabbit will beat a slow turtle every time.

Don't talk to strange bears.

If it looks like a duck,

walks like a duck,

quacks like a duck,

it's probably

an ugly chicken.

Flattery will get you somewhere.

Don't raise your voice to me!

Be quiet!

The End